Energy Science, Engineering and Technology

Energy Science, Engineering and Technology

Nanotechnology Applications in Green Energy Systems
Rajan Kumar, PhD (Editor), Tangellapalli Srinivas, PhD (Editor)
2021. ISBN: 978-1-68507-451-7 (Hardcover)
2022. ISBN: 978-1-68507-479-1 (eBook)

Energy Conversion Systems: An Overview
Sanjeevikumar Padmanaban, PhD (Editor),
Saurabh Mani Tripathi, PhD (Editor)
2021. ISBN: 978-1-53619-131-8 (Hardcover)
2021. ISBN: 978-1-53619-200-1 (eBook)

A Real Alternative for Producing Solar Electricity: The CPV. Current State of CPV: A Review
Leocadio Hontoria, Pedro Pérez-Higueras,
Julio Terrados and Gabino Almonacid (Editors)
2012. ISBN: 978-1-61942-473-9 (Online Book)

Power Systems Applications of Graph Theory
Jizhong Zhu
2011. ISBN: 978-1-60741-364-6 (Hardcover)
2011. ISBN: 978-1-61728-566-0 (eBook)

Silicon Carbide Radiation Detectors
Marzio De Napoli
2013. ISBN: 978-1-61209-600-1 (Hardcover)
2013. ISBN: 978-1-53619-011-3 (eBook)

More information about this series can be found at
https://novapublishers.com/product-category/series/energy-science-engineering-and-technology/

Michael F. Simpson
Editor

The Future of Biodiesel

Copyright © 2022 by Nova Science Publishers, Inc.

All rights reserved. No part of this book may be reproduced, stored in a retrieval system or transmitted in any form or by any means: electronic, electrostatic, magnetic, tape, mechanical photocopying, recording or otherwise without the written permission of the Publisher.

We have partnered with Copyright Clearance Center to make it easy for you to obtain permissions to reuse content from this publication. Simply navigate to this publication's page on Nova's website and locate the "Get Permission" button below the title description. This button is linked directly to the title's permission page on copyright.com. Alternatively, you can visit copyright.com and search by title, ISBN, or ISSN.

For further questions about using the service on copyright.com, please contact:
Copyright Clearance Center
Phone: +1-(978) 750-8400   Fax: +1-(978) 750-4470   E-mail: info@copyright.com.

NOTICE TO THE READER

The Publisher has taken reasonable care in the preparation of this book, but makes no expressed or implied warranty of any kind and assumes no responsibility for any errors or omissions. No liability is assumed for incidental or consequential damages in connection with or arising out of information contained in this book. The Publisher shall not be liable for any special, consequential, or exemplary damages resulting, in whole or in part, from the readers' use of, or reliance upon, this material. Any parts of this book based on government reports are so indicated and copyright is claimed for those parts to the extent applicable to compilations of such works.

Independent verification should be sought for any data, advice or recommendations contained in this book. In addition, no responsibility is assumed by the Publisher for any injury and/or damage to persons or property arising from any methods, products, instructions, ideas or otherwise contained in this publication.

This publication is designed to provide accurate and authoritative information with regard to the subject matter covered herein. It is sold with the clear understanding that the Publisher is not engaged in rendering legal or any other professional services. If legal or any other expert assistance is required, the services of a competent person should be sought. FROM A DECLARATION OF PARTICIPANTS JOINTLY ADOPTED BY A COMMITTEE OF THE AMERICAN BAR ASSOCIATION AND A COMMITTEE OF PUBLISHERS.

Additional color graphics may be available in the e-book version of this book.

Library of Congress Cataloging-in-Publication Data

Names: Simpson, Michael F., editor.
Title: The future of biodiesel / Michael F. Simpson, editor.
Description: New York : Nova Science Publishers, Inc., [2022] | Series:
   Energy science, engineering and technology | Includes bibliographical
   references and index. |
Identifiers: LCCN 2022036092 (print) | LCCN 2022036093 (ebook) | ISBN
   9798886971668 (paperback) | ISBN 9798886971729 (adobe pdf)
Subjects: LCSH: Biodiesel fuels.
Classification: LCC TP359.B46 F88 2022 (print) | LCC TP359.B46 (ebook) |
   DDC 665/.37--dc23/eng/20220808
LC record available at https://lccn.loc.gov/2022036092
LC ebook record available at https://lccn.loc.gov/2022036093

Published by Nova Science Publishers, Inc. † New York

Contents

Preface vii

Chapter 1   **Concurrent Biodiesel Production and Wastewater Remediation Using Microalgae** 1
P. J. Welz and A. Ranjan

Chapter 2   **Are Microreactors the Future of Biodiesel Synthesis?** 47
Rosilene A. Welter, João L. Silva Jr., Marcos R. P. de Sousa, Mariana G. M. Lopes, Osvaldir P. Taranto and Harrson S. Santana

Chapter 3   **Effect of Biodiesels - Bioethanol Fuel Mixture on Performance, Characteristics in Diesel Engines** 83
M. Acaroğlu and H. Köse

Chapter 4   **The Study of Properties in Biodiesel/Butanol and Biodiesel/Diesel/Butanol Blends** 111
S. D. Romano

Chapter 5   **Natural Attenuation of Soil with Biodiesel and Its Potential Ecotoxicological Impacts** 127
Guilherme Dilarri, Carolina Rosai Mendes, Vinícius de Moraes Ruy Sapata, Ivo Shodji Tamada, Paulo Renato Matos Lopes, Renato Nallin Montagnolli and Ederio Dino Bidoia

Index 137

Preface

Biodiesel is a renewable and biodegradable form of diesel derived from vegetable oils, animal fats, or recycled grease. Biodiesel offers a number of benefits compared to standard diesel, including environmental and safety benefits. This book includes five chapters that explain various aspects of biodiesel. Chapter One focuses on various aspects of biodiesel production from microalgal species that have the potential to remediate different types of wastewaters. Chapter Two aims to show whether microreactors can replace conventional biodiesel production processes and how this replacement technology could be carried out. Chapter Three studies the effects of Cynara biodiesel-bioethanol-diesel fuel mixtures on engine performance characteristics. Chapter Four presents a study of flash point and refractive index in biodiesel/butanol and biodiesel/diesel/butonal blends. Lastly, Chapter Five presents how microbial biodegradation may ultimately affect the overall degradation of hydrocarbons, including biodiesel.

Chapter 1 - Phycoremediation of industrial or municipal effluent using microalgae reduces the organic and/or inorganic loads of the wastewater while simultaneously supporting the growth of a valuable form of biomass that can be exploited as an energy carrier. Such integration involves recycling, reuse and remediation of water and nutrients for algal growth, thereby providing an eco-friendly and sustainable means for biofuel production in support of a circular economy. This chapter focusses on various aspects of biodiesel production from microalgal species that have the potential to remediate different types of wastewaters. The discussions include the status quo and future trends in microalgal wastewater treatment, descriptions of oleaginous algal species used for biodiesel production and other applications, and techniques used for harvesting and converting algal bio-oil to biodiesel.

Chapter 2 - Microfluidic devices or microdevices refer to systems with a characteristic length in the micrometer range. Systems in this size allow handling small quantities of reagents and samples, with reduced residence time, better control of chemical species concentration, high heat and mass transfers, and high surface/volume ratio. These characteristics led to the application of these microdevices in several areas, such as biological systems, energy, liquid-liquid extraction, food, agricultural sectors, pharmaceuticals, flow chemistry, microreactors, and biodiesel synthesis. Microreactors are devices that have interconnected microchannels, in which small amounts of reagents are manipulated and react for a certain period of time. The traditional characteristics of microreactors are less quantities of reagents and samples, high surface area in relation to volume (10000 m2 m-3), reduction of resistance to heat and mass transfer, reduced reaction times, and narrower residence time distributions. In recent years, several studies have been carried out on biodiesel production in microreactors that explore the influence of operating conditions, mixing and reaction yield, numbering, and especially the microdevices design. Despite all the advantages of microreactors, the literature shows that there are only a few applications on an industrial scale. Two main reasons that hinder the adoption of this technology are the scale-up to a large enough volume to deliver the necessary production capacity and the costs related to industrial manufacturing microreactors. It is often stated that large-scale production of microreactors can be easily achieved by numbering-up. However, research shows that an incredibly high number of microdevices would be needed, which results in a technical unfeasibility and a strong impact on the construction costs of the industrial system. The present review aims to show whether microreactors can replace conventional biodiesel production processes and how this replacement technology could be carried out. The current chapter was divided into the following sections: Introduction, Synthesis and Purification of Biodiesel in Microreactors, Fundamentals of CFD, and Fundamentals of Scale-up. Finally, conclusions and future perspectives are exposed.

Chapter 3 - Cynara cardunculus is a Mediterranean perennial herb. Cynara cardunculus is a herbaceous, robust, long-lived plant that can be grown even in non-agricultural lands in Turkey, with low water requirement, high yield. In this study, the effects of Cynara biodiesel-bioethanol-diesel fuel mixtures on engine performance characteristics (engine power, torque, emissions) were investigated. Cynara cardunculus biodiesel was mixed with different proportions of bioethanol diesel fuel. It has been named as diesel (D100), biodiesel (B100) and diesel-biodiesel-bioethanol (B5E5-B7E5).

According to the experimental results, maximum engine power (44.44 kW) and maximum torque (166 Nm) B7E5, the highest thermal efficiency was acquired with B7E5 fuel mixture at 3000 rpm. The lowest specific fuel consumption (be) was measured with D100 fuel at 2000 rpm as 218.18 g/kWh.

Maximum in-cylinder pressure was measured as 103.2 bar B5E5 at 3000 rpm. Max heat release rates (HRR) were obtained with B7E5 at 2000 rpm and 3000 rpm as 524 kJ/m3.deg and 602.4 kJ/m3.deg, respectively. CO emission values for B100, B5E5 and B7E5 fuel mixtures decreased averagely 82.84%, 42.975% and 52.73% compared to standard diesel operation. CO_2 emissions increased averagely 2.1%, 1.8%, 3.04%. HC emissions significantly reduced. NOx emissions for all fuel increased. Consequently, cynara cardunculus seed oil biodiesel and bioethanol can be mixed at certain rates at diesel engine.

Chapter 4 - Biodiesel/butanol and biodiesel/diesel/butanol blends have recently received international attention due to the improvement in their properties because of the addition of butanol to the fuel/s, and the possibility of massive use of such systems in a near future. The data about these blends in the scientific literature are mainly related to the study of fuel consumption, combustion performance, and exhaust emissions (NOx, CO, HC, and smoke) in diesel engines. However, most studies also include the determination of some properties of technological interest in a few samples.

This chapter presents the study of two important properties - flash point (related to safety during transportation and storage of the blends) and refractive index (an optical property of translucent substances that can give an account of their quality) - in biodiesel/butanol and biodiesel/diesel/butanol blends. The flash point has been very poorly studied in these systems, whereas the refractive index has not been studied.

Experimental results show that, at low butanol concentrations in the blends, the flash point values decrease sharply, whereas above 20% butanol content in volume in the blends, they remain practically constant. Refractive index values decrease linearly with increasing butanol content in the blends and temperature.

Chapter 5 - Petroleum and its derivatives cause numerous toxicity-related impacts on the environment. The evaluation of such impacts can promote adequate treatments of oil-based residues. A better understanding of the degradation of hydrocarbons aids researchers to predict the environmental impacts and the biodegradation processes that yield decreased toxicity levels. However, petroleum derivatives are not the sole matter of concern. Another important discussion targeting the environmental behavior of biodiesel should be brought up. Soil microorganisms may improve the biodegradation of hydrocarbons due to their diversity and robust enzymatic apparatus during natural attenuation processes. Still, some compounds can have their negative impacts on biota evaluated according to three parameters: ecotoxicological potential in soil (i), specific biodegradation processes (ii), and post-treatment bioassays (iii). This review presents how microbial biodegradation may ultimately affect the overall degradation of hydrocarbons.

Chapter 1

Concurrent Biodiesel Production and Wastewater Remediation Using Microalgae

P. J. Welz and A. Ranjan
Cape Peninsula University of Technology,
Bellville, Cape Town, South Africa

Abstract

Phycoremediation of industrial or municipal effluent using microalgae reduces the organic and/or inorganic loads of the wastewater while simultaneously supporting the growth of a valuable form of biomass that can be exploited as an energy carrier. Such integration involves recycling, reuse and remediation of water and nutrients for algal growth, thereby providing an eco-friendly and sustainable means for biofuel production in support of a circular economy.

This chapter focuses on various aspects of biodiesel production from microalgal species that have the potential to remediate different types of wastewaters. The discussions include the status quo and future trends in microalgal wastewater treatment, descriptions of oleaginous algal species used for biodiesel production and other applications, and techniques used for harvesting and converting algal bio-oil to biodiesel.

Keywords: phycoremediation, microalgae, wastewater, biodiesel

In: The Future of Biodiesel
Editor: Michael F. Simpson
ISBN: 979-8-88697-166-8
© 2022 Nova Science Publishers, Inc.

Introduction

Microalgae can remove N, P and organic compounds (macronutrients) and potentially toxic contaminants from urban and industrial wastewaters in a process known as phytoremediation. Algal phytoremediation is reliant on autotrophic and/or heterotrophic energy-generating metabolic pathways. Autotrophic algae capture CO_2 via photosynthesis to form organic molecules intracellularly, while heterotrophic algae rely on external organic carbon substrates. The microalgal species that play functional roles in wastewater treatment/remediation may be: (i) strictly autotrophic, (ii) strictly heterotrophic, or (iii) mixotrophic. Mixotrophic algae have the ability to shift between autotrophic and heterotrophic growth, with the primary metabolic switch being the presence or absence of sufficient light (Gupta et al., 2019). It has recently been highlighted that significant amounts of the greenhouse gas N_2O are emitted from autotrophic microalgal-based wastewater treatment systems treating N-rich effluents (Casagli et al., 2021; Plouviez and Guieysse, 2020).

Waste stabilization ponds are simple algal ponds that have been used for decades to passively remediate various types of wastewaters. For urban wastewater, they function by removing macronutrients. Organic molecules are metabolized (removed) by heterotrophs and formed (added) by autotrophs. This can lead to relative diurnal increases and nocturnal decreases in the concentration of organics, typically measured using the chemical oxygen demand (COD) test (Paddock et al., 2020). Although cost-effective, traditional algal ponds are not as efficient as conventional secondary wastewater treatment processes such as the activated sludge process and are therefore not suitable for remediating large volumes of effluent. In larger urban environments they are often employed for 'polishing' *viz.* as a tertiary (final) treatment step to remove excess N and P from secondary effluent before discharge.

In line with the global trend towards bio-circular economies, microalgal phytoremediation has become more holistic. This has been ignited by the fact that microalgae can treat wastewaters while simultaneously forming biomass that can be harvested and utilized for various low to high value products. For example, the biomass can used after minimal processing as animal feeds or biofertilizers or for basic nutrient recovery (Rearte et al., 2021). Alternatively, valuable pigments can be extracted from some algal species, or extracts (lipids, carbohydrates) can be used as substrates for biofuels or other bio-based products, including bioplastics. Along with the establishment of algal wastewater biorefineries, there is a move toward the introduction of more sophisticated algal wastewater treatment systems, including high-rate algal ponds (HRAPs), photobioreactors (PBRs) and hybrid HRAP-PBR systems for concurrent wastewater remediation and algal biomass generation. In terms of wastewater treatment efficiency, the use of microalgal species that have been harvested from facilities treating wastewater may result in more stable process control. This is because these native microalgal consortia (NMC) have already been naturally selected according to the specific wastewater characteristics and (in the case of outdoor systems), to the prevailing (ambient) temperatures and light intensities (Moreno-García et al., 2021).

Wastewater Treatment Using Microalgae: Background

High-Rate Algal Ponds and Photobioreactors

High-rate algal ponds, mostly in the form of 'raceway' ponds have been widely introduced for concurrent wastewater remediation and biomass harvesting. The wastewater in HRAPs is mechanically circulated and the systems are open to the environment, relying on natural light for the growth of autotrophic algal species. The systems are inevitably heterotrophic/mixotrophic to some extent as completely autotrophic growth requires constant light. A comprehensive review of literature suggests that, on average, high microalgal yields and efficient COD and N removal can be achieved with loading rates of around 200 gCOD m^{-3}/day and 20 gTN m^{-3}/day (Torres-Franco et al., 2021). Apart from nutrient removal, it has also been shown that biodegradation of pollutants such pharmaceuticals takes place in HRAPs (García-Galán et al., 2020)

In all algal wastewater treatment systems, including HRAPs, the light intensity, wavelength/s and photoperiod, pH, temperature, and N and P concentrations all have effects on biomass speciation and productivity, which typically go hand-in-hand with wastewater treatment efficiency (Arcial et al., 2021; Singh and Mishra, 2021;). Systems that require autotrophic microalgal growth should ideally be designed and located to allow good light penetration and maintain relatively stable temperatures. Such HRAPs tend to have larger spatial footprints than conventional wastewater treatment systems because they need to be shallow to allow sufficient light penetration. Penetration of light in HRAPs is hampered by the presence of suspended solids (SS), a phenomenon known as shading. The algae themselves can contriubute to shading, which is a function of both the concentration and type of SS and the depth of the systems (Sorooth et al., 2022). It may be countered to some degree by pre-treatment to reduce the concentrations of SS. This has the added advantage of reducing the numbers of other microbes in the wastewater that may compete with the microalgae for substrates or predate on the algae (Gupta et al., 2019). However, there is a school of thought that for organic wastewaters, mixotrophic algal systems with significant heterotrophic microalgal metabolism are the way of the future because they are only partially reliant on light and also have smaller spatial footprints. Indeed, large-scale deep water mixotrophic HRAPs containing relevant algal consortia have shown promising results over long periods (Watanabe & Isdepsky, 2021).

Photobioreactors (PBRs) are typically closed, more technologically advanced systems that are usually artificially heated, aerated and illuminated. However, PBRs can also be located outdoors and/or open to their environment where they are reliant on ambient conditions for growth (Garcia et al., 2017). The aeration gas mixture in PBRs may be supplemented with additional CO_2, depending on autotrophic growth requirements. Aeration also assists with mixing in PBRs. The ability to control temperature and light allows indoor PBRs to be operated to encourage autotrophic, heterotrophic or mixotrophic growth. The latter is typically induced by alternating dark and light conditions. Depending on the algal species that are present, it is possible to achieve high nutrient removal rates by optimizing mixotrophic growth (Ding et al., 2021).Operation of indoor PBRs is not economically viable for treatment of large volumes of effluent, particularly if the downstream algal-based products are not high in value. Much of the research based on the use of PBRs has been directed towards growing single microalgal species or selected algal consortia, often requiring sterile wastewater that is unrealistic to achieve at scale. In contrast, HRAPs and outdoor PBRs are widely used with the acknowledgement that microalgal growth is slow in colder climates, that species control is limited so that mixed consortia of microalgae, bacteria and fungi is almost inevitable (Singh and Mishra, 2021).

Hybrid Systems and Other Novel Systems

Hybrid systems that consist of HRAPs that are fed with algal cultures from PBRs are becoming popular in cases where microalgal species with relevant characteristics for downstream applications are required. The rationale is that these systems maximize the growth of the desirable species while minimizing the costs associated with the operation of PBRs alone. Other novel technologies include bioelectrical systems and microalgal membrane bioreactors. Microbial fuel cells are gaining interest in the field of electricity generation with concurrent wastewater remediation. For example, an electrogenic strain of *Coelastrella* sp. was able to generate 314 mW m^{-2} in a lab-scale raceway HRAP (Raja et al., 2022). Although there are still many challenges, and the amounts of electricity that can be generated are low, constant improvements may very well see this technology widely applied in future. In the case of algal membrane bioreactors, exopolymeric substance accumulation may lead to excessive membrane fouling (Onyshchenko et al., 2020), although it has been demonstrated at lab scale that aeration can be used to improve overall performance by circulating the biomass and providing O_2 (Ding et al., 2020). The aeration costs may be offset to some extent by the improved ease of biomass harvesting with membranes.

Microalgal-Bacterial Wastewater Treatment Systems

The concept of wastewater remediation with a combination of microalgae and bacteria is not new, and microalgae-bacteria (MAB) systems have existed for decades (García-Galán et al., 2020). These are based on the premise that autotrophic algae generate O_2 in the presence of light, reducing aeration costs and providing O_2 required for nutrient removal by certain functional bacterial taxa. Notably, for wastewater with high concentrations of environmentally toxic NH_3/NH_4^+ such as urban effluent, O_2 is required by (chemo)autotrophic ammonium oxidizing bacteria for the conversion of NH_3/NH_4^+ to less toxic NO_2^- and NO_3^- in a process known as nitrification. The autotrophic algae, in turn, utilize the CO_2 generated by heterotrophic bacteria for photosynthesis (Flores-Salgado et al., 2021). A study by Soroosh et al. (2022) determined that a ratio of algae to heterotrophic bacteria of 1.64 was required to maintain O_2 consumption and generation at an equilibrium. Apart from heterotrophic bacteria, mixotrophic algae in the absence of light and heterotrophic algae may be the largest O_2 consumers and the most metabolically active microbes in terms of organic substrate utilization in MAB systems (Flores-Salgado et al., 2021). In cases where algal growth is paramount, a way of controlling the overgrowth of bacteria may be to alternate the feed from high C/N to low C/N under strictly heterotrophic (dark) conditions. With synthetic feed, researchers have shown that this can reduce the bacterial:microalgal ratio. In the case of a simulated MAB system containing defined artificial media for growth of *Tetradesmus obliquus* (formerly *Scenedesmus obliquus*), researchers were able to reduce the ratio from 1.6 to 0.03 (Di Caprio et al., 2019). However, this may be difficult to achieve with 'real' wastewater as demonstrated with diluted (10%) olive mill wastewater permeate. Another advantage of MAB systems is that the biomass tends to settle passively, while in pure algal systems the biomass may flotate because unicellular microalgal species tend to have negative charges on their surfaces (Singh and Mishra, 2021; Soroosh et al., 2022). With MAB systems, ammonium oxidizing bacteria (AOB) may compete with algae for NH_3, especially when the light intensity and O_2 concentration are low (González-Camejo et al., 2022) and there is a small risk that some bacterial secondary metabolites may be algicidal (Sanchez-Zurano et al., 2021).

Algal-Biodiesel Wastewater Biorefineries (ABWBs)

Introduction

In addition to urban wastewaters and other effluents that are amenable to algal growth, even recalcitrant wastewaters such as anaerobic digester effluents, textile wastewater, pulp and paper industry wastewater and winery wastewater show promise for use in algal biodiesel wastewater biorefineries (ABWBs) (Almaguer et al., 2021). In many cases, some form of primary or secondary pre-treatment is applied, while in other cases raw wastewater is used. Apart from the bioreactor type and operational considerations, the achievable lipid concentration in the microalgal biomass is a critical factor to consider for ABWBs. A wide range of lipid concentrations can be attained, depending on both the wastewater type and the microalgal strain/s. In a comprehensive literature review, El-Sheekh et al. (2021) reported lipid concentrations in *C. vulgaris*-rich biomass ranging from 9% in aqua culture effluent to 58% in piggery effluent, with the highest concentration (64%) being reported in *T. obliquus* grown in brewery effluent. The costs of ABWBs include expenses for plant installation and operation, biomass harvesting, and lipid extraction and conversion to biofuel. A recent techno-economic assessment of different strategies estimated that the cost of biofuels from ABWBs ranged from 0.14 to 24.33 USD/L (Aggarwal & Remya, 2021). The cost bottleneck is usually the harvesting step, which can account for up to 30% of the production costs of algal-based biodiesel (Kadir et al., 2018). The input and operational costs are, however, offset by the costs of traditional wastewater treatment and, if applicable, carbon and environmental credits (Aggarwal & Remya, 2021).

Wastewater Treatment Systems for Growth of Lipidogenic Biomass

In line with the advent of circular economy principles, there have been hundreds of publications on growth of microalgae in various wastewaters for downstream value-added applications. Many full-scale HRAPs for biofuel generation are already operating successfully (García-Galán et al., 2020; García-Galán et al., 2020). By comparing the results obtained from a number of studies, it appears that the size of HRAPs is negatively correlated with microalgal productivity and biomass lipid content. This was confirmed by a group of researchers who found a significant difference in algal productivity in three systems of different sizes (surface areas of 5 m^2, 330 m^2 and 10000 m^2) fed with the same influent (urban pond effluent) at the same horizontal velocity (0.2 m^{-2}) and hydraulic retention time (HRT) of 8 days, in the same geographical location (Sutherland et al., 2020). Lower algal productivity in larger systems appears to mainly be a function of inadequate mixing because effective mixing is easier to achieve in smaller systems, while laminar flows and dead zones are more likely to occur in larger HRAPs (Sutherland et al., 2020).

Photobioreactors
There are many studies detailing the growth of lipidogenic microalgal species in aerated and artificially illuminated PBRs with a variety of wastewaters. The most recent of these (SCPOUS search 2019-2022) were compared in terms of algal species, selected operational parameters (excluding aeration), and the type of wastewater. In all cases, concurrent wastewater remediation was successful and removal rates and the specific wastewater compositions are not discussed in any detail in this chapter. Table 1 contains information on recent studies conducted using PBRs for cultivation of microalgal species under different light conditions. Those promoting autotrophic algal growth are operated under continuous light, while heterotrophic and autotrophic growth is promoted using light:dark cycles, the most common being 16 hrs light and 8 hrs dark. In an effort to maintain pure or dominant growth of selected microalgal inocula, many of these studies made use of pre-sterilized wastewater, or centrifuged supernatant. These practices are not economically

viable at scale for production of a relatively low-value product such as biodiesel, especially in systems that already require illumination and/or aeration. In addition, wastewater is commonly diluted in order to provide optimal nutrient ratios and/or decrease the concentration of microalgal growth inhibitors. For example, dilution of winery wastewater (50%) was found to reduce agal stress and increase lipid production in *Arthrospira platensis and C. vulgaris* (Spennati et al., 2020). Even with dilution, the wastewater may not be sufficiently remediated, requiring further treatment before discharge (Marchão et al., 2021). In addition, there are questions as to the sustainability of wastewater dilution, especially in water-stressed countries. For these reasons, there has been limited application of high-technology PBRs for biodiesel generation at industrial scale. Nonetheless, these small-scale studies provide fundamental insight into microalgal growth using wastewater that may be utilized when selecting species for HRAPs and other systems that are more economically viable to operate.

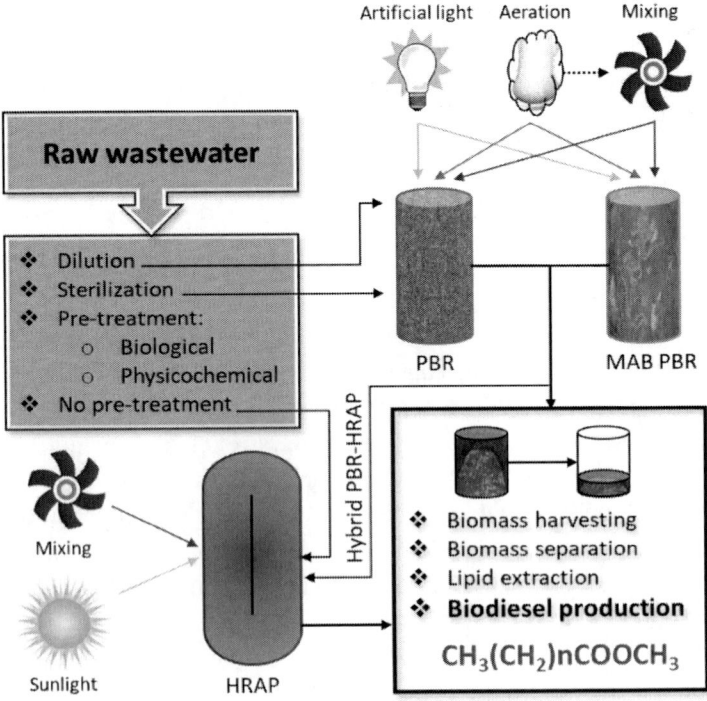

Figure 1. Schematic flow chart depicting the most typical inputs and processes for experimental and full-scale algal biodiesel wastewater biorefineries.

Table 1. Photobioreactors for cultivation of lipidogenic microalgal species in wastewater (SCOPUS 2019-2022)

Microalgal species	Selected operational parameters	Wastewater type	Reference
Continuous light			
Tribonema sp.	Volume: 150 mL 9 days in batch mode	Petrochemical	Huo et al. 2019
Chlorella. vulgaris	Volume: NG 12 days in batch mode	Dairy: Sterilized & diluted	Khalaji et al. 2021
Arthrospira maxima, Tetradesmus obliquus, C. vulgaris, Auxenochlorella protothecoides	110 mL 8 days in batch mode	Winery: Sterilized & diluted	Marchão et al. 2021
Arthrospira platensis, C. vulgaris	200 mL 21 days in batch mode	Winery: Diluted	Spennati et al. 2020
Scenedesmus abundans, C. vulgaris, Nannochloropsis salina	500 mL 30 days SC at 2-day HRT	Urban: RO concentrate	Mohseni et al. 2021b
Scenedesmus, Chlorella spp.	Volume: NG 2-stage: 25 & 15 day in batch mode	Palm oil mill: Diluted	Hariz et al. 2019
Chlorococcum robustum AY122332.1	3120 mL/continuous at 2-day HRT	Rare earth element tailings: plus Na_2CO_3	Geng et al. 2022
Alternating light:dark			
C. vulgaris, Botryococcus braunii, Ankistrodesmus falcatus, T. obliquus	Volume: 800 mL / 10 days in batch mode Light:dark 16:8 hrs	Urban: Tertiary, filter sterilized	Lavrinovičs et al. 2021
C. vulgaris	Volume: 500 mL / 15 days batch mode Light:dark 12:12 hrs	Textile	Bellucci et al. 2021
T. obliquus, Asterarcys, Quadricellulare, Desmodesmus, Pseudopediastrum spp.	Volume: 500 mL / 11 days batch mode Light:dark 16:8 hrs	Urban: Settled, filter sterilized	Do et al. 2019
Chlorococcum, Stigeoclonium spp.	Volume: 1000 mL / SC 1.7 to 6-day HRT Light:dark 16:8 hrs	Slaughterhouse: Settled	Rearte et al. 2021
C. vulgaris, Chlorella, Scenedesmus spp.	Volume: 50 mL / 8 days batch mode Light:dark 14:10 hrs	Piggery: Centrifuged supernatant	Liu et al. 2022
Microalgal species	Selected operational parameters	Wastewater type	Reference
Scenedesmus quadricauda, Tetraselmis suecica	Volume: 1000 mL / 12-day cycles Light:dark 12:12 days (2 stage)	Dairy	Daneshvar et al. 2019

Table 1. (Continued)

Microalgal species	Selected operational parameters	Wastewater type	Reference
Natural light			
Chlorella, Scenedesmus spp.	Volume 45 000 mL (pilot) 7 days batch mode	Winery: Tertiary effluent	Avila et al. 2022

Tetradesmus obliquus formerly known *as Scenedesmus obliquus* HRT = hydraulic retention time SC = semi-continuouse.

Some of the more novel studies include the growth of the filamentous *Tribonema* sp. in petrochemical wastewater, where a range of organic molecules were successfully biodegraded (Huo et al., 2019), and growth of a native *Chlorococcum* sp. on high N rare earth element tailings wastewater supplemented with Na_2CO_3 as a source of C (Geng et al., 2022).

Microalgal Bacterial Photobioreactors

In contrast to PBRs for growth of only selected microalgal species, the use of MAB PBRs (Table 2) that do not require sterile wastewater and allow growth of microbes present in the wastewater is more realistic as they do not require the use of sterile effluent.

Raw, undiluted effluent can be treated using these systems which tend to be larger than non-MAB PBR experimental systems and are often operated in continuous mode, mimicking full-scale systems. Indeed, stable system performance has been demonstrated in a large scale (1.6 kL) outdoor cascading thin layer MAB PBR system. In this study, although there were seasonal fluctuations in the bacterial species composition, the growth of the selected *Scenedesmus* strain was relatively stable (Sánchez Zurano et al., 2020).

Good microalgal growth can be achieved in MAB PBRs, with high lipid concentrations and profiles for biodiesel generation (Marizzi et al., 2020). However, control of microalgal:bacterial stability is challenging, and intermittent bacterial overgrowth and fluctuating COD removal efficiencies can occur, as demonstrated with growth of *Chlorella sorkiniana* in anaerobic digestate (Paddock et al., 2020).

Table 2. Microalgal bacterial photobioreactors systems for cultivation of lipodogenic microalgal species in wastewater (SCOPUS 2019-2022)

Microalgal species	Selected operational parameters	Wastewater type	Reference
Chlorella sorokiniana	Volume: 200 mL / 7 days batch mode Light:dark 16:8 hrs	Urban anaerobic digestate	Paddock et al. 2020
Tetradesmus, Scenedesmus spp.,	Volume 1000 mL / 36 hrs batch mode Light: 24 hrs	Tannery	Moreno-García et al. 2021
Scenedesmus acuminatus, Chlorella, Scenedesmus, Chlamydomonas spp.	Volume: 2500 mL / 70 days continuous mode at 7-day HRT Light:dark 12:12 hrs	Milk whey processing	Marazzi et al. 2020
Chlorella, Coelastrella spp.	Volume: 500 mL / 8 days batch mode Light: 24 hrs	Piggery	Qu et al, 2021
C. sorokiniana	Volume: 2000 mL / 30 days sequencing batch mode Light:dark 16:8 hrs	Urban	Kotoula et al. 2020
C. vulgaris, Chlamydomonas sp., *Chlorella kessieri, Scenedesmus acutus*	Volume: 3000 mL / 225 days continuous mode at 10.6-day HRT Light dark: 12:12 hrs	Piggery: Diluted (5%)	García et al. 2019
Scenedesmus sp.	Volume: 1600000 mL / 300 days semi-continuous mode at 0.3 day^{-1} Sunlight	Urban	Sánchez Zurano et al. 2020
Desmodesmus sp.	Volume: 100 mL / 7 days batch Light: 24 hrs	Piggery: Digestate & anoxic effluent	Wang et al. 2020
C. sorokiniana	Volume: 3000 mL / 10 days batch Light:dark 16:8 hrs	Dairy	Makut et al, 2019

HRT = hydraulic retention time.

Hybrid Systems

For ABWBs, good growth of lipidogenic microalgal species is critical in ensuring that the entire process of biodiesel production is economically viable. Sidestream bioreactors with defined media have been used to grow lipidogenic microalgal species and consortia suitable for biodiesel production, including *Scenedesmus almeriensis* (Sanchez-Zurano et al., 2021), *Halamphora coffeaeformis* (Martín et al., 2016), *C. vulgaris* (Heidari et al., 2016), *Tetraselmis* sp. (Narala et al., 2016), *Chlorella ellipsoidea, Chlorococcum infusionum* (Satpati et al., 2016) and *Chlorella pyrenoidosa* (Siddiqui & Suneetha, 2019). Hybrid PBR-HRAP systems are suited to ABWBs because it is possible to achieve a measure of control over which microalgae dominate in the HRAP. Examples include the growth of *S. almeriensis* in a 32 m² thin-layer cascade PBR feeding a HRAP treating urban effluent in semi-continuous mode (Sanchez-Zurano et al., 2020), and rudimentary outdoor PBRs consisting of a series of 500 L see-through tanks for growth of *C. vulgaris* and *Chlorella protothecoides* used to feed a waste stabilisation pond, also treating urban effluent (Oberholster et al., 2021).

A comprehensive lifecycle assessment comparing the environmental impact of biodiesel production from a hybrid mixotrophic PBR-HRAP system with that of fossil-based biodiesel found 76% and 75% savings in global warming potential and fossil energy requirements, respectively. The assessment was based on growth of *C. vulgaris* in defined culture media (Adesanya et al., 2014). Microalgal lipid production in raceways is optimal in starvation (nutrient deficient) conditions (Adesanya et al., 2014; Heidari et al., 2016; Narala et al., 2016) which can easily be achieved using artificial media, but not necessarily with wastewater. It is therefore difficult to predict whether ABWBs would have a greater or lesser environmental impact than biodiesel biorefineries operated using defined media, although the cost of using defined media renders the process uneconomical.

Wastewater Pre-Treatment Strategies

Studies have focused on the use of biological or physicochemical pre-treatments to enhance microalgal productivity, lipid production and wastewater remediation in PBRs, some of which are provided in Table 3. Examples of 2-stage biological systems consisting of biological pre-treatment followed by PBR cultivation include anaerobic acidogenic fermentation (You et al., 2021; Zhang et al., 2021), pre-digestion using an upflow anaerobic sludge blanket reactor (de Souza Leite et al., 2019), an anaerobic moving bed bioreactor (Zkeri et al., 2021), and an anaerobic membrane bioreactor (Gao et al., 2020). As well as improving the C:N ratio for microalgal growth and lipid production, the anaerobic processes produce VOAs and solubilize particulates to release NH_3/NH_4^+ and P as microalgal substrates (You et al., 2021; Zhang et al., 2021; Gao et al., 2020).

Ozonation (Hu et al., 2020, 2021), chlorination (Hu et al., 2021), magnetic field application (Feng et al., 2020) and acid precipitation (Hilares et al., 2021) are examples of physicochemical pre-treatments that have been used to stimulate microalgal growth and lipid production. Apart from reducing the bacterial counts in some instances, in most cases the exact mechanisms involved in enhancing microalgal growth have not been comprehensively elucidated.

Table 3. Physicochemical or biological pre-treatment methods for cultivation of lipodogenic microalgal species in wastewater (SCOPUS 2019-2022)

Microalgal species	Pre-treatment	Selected operational parameters	Wastewater type	Reference
C. vulgaris	AnAF	1000 mL PBR / 11 days batch mode Light:dark 12:12 hrs	Mariculture: Diluted	You et al. 2021
C. sorokiniana	UASB 3-day HRT	50000 mL PBR /7 days batch mode Light:dark 16:8 hrs	Urban & Piggery	de Souza Leite et al. 2019
T. obliquus, C. vulgaris, C. sorokiniana	Ozone or chlorine	Alginate immobilized MAB PBR / 7 days batch mode Light:dark 16:8 hrs	Meat processing: Settled & DAF treated	Hu et al., 2021 Hu et al., 2020
Chlorella pyreoidsa	Magnetic field	5000 mL PBR / 7 days batch mode Light:dark 12:12	Urban: Centrifuged supernatant	Feng et al. 2020
C. sorkiniana	AnMBBR	2000 mL MAB PBR / 7 days batch mode Light:dark 16:8 hrs	Dairy: Sieved	Zkeri et al. 2021
C. vulgaris	Acid ppt.	5000 mL MAB PBR / continuous 3.4-day HRT Light: 24 hrs	Poultry: Coarse filtered	Hilares et al. 2021
C. pyrenoidsa	AnMBR	1000 mL membrane PBR / continuous 4-day HRT Light: 24 hrs	Urban: Filtered (0.22 um)	Gao et al. 2020
C. vulgaris	AnAF	800 mL MAB PBR / 10 days batch mode Light: dark 12:12 hrs	Mariculture: Diluted (10%) & centrifuged	Zhang et al. 2021

AnAF = anaerobic acidogenic fermentation PBR = photobioreactor UASB = upflow anaerobic sludge blanket A/O = anaerobic oxic process AnMBBR = anerobic moving bed biofilm reactor HRT = hydraulic retention time.

Potential for Enhanced Growth of Lipidogenic Microalgal Species in Wastewater

Green algae and cyanobacteria are the two major photosynthetic organisms that have been employed to produce liquid biofuels like bio-alcohols and biodiesel. Microalgae, as the name implies, are algae that cannot be seen with the naked eye as individual cells. They are abundantly available in nature and many of them have the capacity to synthesize lipids, *viz.* they are lipidogenic. Microalgal lipids can be synthesized using either carbohydrates or acetate as substrates, and lipid accumulation is an energy storage process that involves utilization of ATP via photophosphorylation and oxidative phosphorylation. Oleaginous algal species are those that accumulate lipids to ultimately make up more than a fifth of their biomass. Microalgal lipids can be polar and/or non-polar. Polar lipid fractions are comprised mainly of glycolipids and phospholipids which are polyunsaturated fatty acids (PUFA) and are not convertible into biodiesel, while triacylglycerides (TAG) are the non-polar fractions that can be further transformed into biodiesel through transesterification. Under stress conditions, microalgal cells produce intracellular TAG as non-structural storage lipids. Strains that have the ability to synthesize fatty acids that result in the most appropriate forms of TAG are most suited to the production of third generation biodiesels. Oleic acid and palmitic acid have been reported as the major fatty acids produced by *A. platensis*, while elaidic acid has been reported as the dominant fatty acid produced by *Nannochloropsis* sp. and *Botryococcus braunii* (Pradana et al., 2017). However, variations in the nutritional and physical environments of microalgal cells translate into variations in the fatty acid composition of TAGs (Xu and Beardall, 1997; Sushchik et al., 2003).

Many microalgal species are capable of tolerating and growing in extreme wastewater environments such those with high alkalinity, and/or high COD, and/or high salinity and fluctuating nutrient supplies (Salih, 2011; Razzak et al., 2013; Cheirsilp, Thawechai and Prasertsan, 2017). Promising lipidogenic microalgal species that have been successfully grown in wastewater include *T. obliquus, C. vulgaris, C. protothecoides, Chlorella marina, Ettlia oleoabundans* (previously known as *Neochloris oleoabundans*), *Tetradesmus bernardii, Ankistrodesmus* sp, *Dunaliella tertiolecta, Chlorella sorokiniana, Nannochloropsis oculata, Haematococcus pluvialis, Thalassiosira pseudonana, Isochrysis galbana, Golenkinia* sp., *Monoraphidium* sp., *Coelastrella* sp., and *Tetraselmis* sp. (Zhang et al., 2014; Gutierrez et al., 2016; Cheirsilp et al., 2017; Liu et al., 2017; Daneshvar et al., 2018; Matich et al., 2018; Ganeshkumar et al., 2018; Gao et al., 2018; Ling et al., 2019; Swain et al., 2020; Ye et al., 2020; Ren et al., 2021; Wang et al., 2021; Zhang et al. 2021; Han et al., 2021; Liu et al., 2021a; Liu et al., 2021b; Lee et al., 2021; Nambukrishnan and Singaram, 2022; Yang et al., 2022; Zhao et al., 2022; Kirchner et al., 2022). Some of the multicellular cyanobacterial species, such as *Oscillatoria* spp., *Sargassum muticum, Synechocystis* spp., *Spirulina* spp.,

Nostoc spp., *Turbinaria turbinata*, *Caulerpa scalpelliformis*, *Pseudanabaena* sp., *Anacystis* sp., *Aphanothece flocculosa*, *Gracilaria edulis*, and *Lyngbya majuscule* also show impressive metal sorption properties (Kulal et al., 2020).

Table 4. Culture conditions, biomass and lipid yields of a selection of oleaginous microalgal species in wastewater

Microalgal species & wastewater	Culture conditions				Biomass yield (g/L)	Lipid yield (g/L)	Reference
	Temp. (°C)	Time (days)	Aeration	Light			
C. pyrenoidsa Urban and distillery effuent (1:1)	30	7	Agitation @ 140 rpm	2.93 W/m² @ 2000 lux light:dark 12:12 hrs	NG	3	Ling et al. 2019
Immobilised *Nannochloropsis* sp. Secondary palm oil mill effluent	30±2	7	10% CO_2 @ 0.4-1.0 vvm	47 µmol photon/m².s⁻¹ Light:dark 24:0 hrs	1.30 ±0.05	0.36 ±0.10	Cheirsilp et al. 2017
T. obliquus Urban (settled raw wastewater)	23.2±2	10	Agitation @ 120 rpm	2500-3000 lux Light:dark 24:0 hrs	1.35	0.38	Han et al. 2021
Chlorella marina Secondary tannery effluent	27	9	8% CO_2 @ 100 L/hr	200 µmol photon/m².s⁻¹ Light:dark 12:12 hrs	3.84	1.56	Nambukrishnan & Singaram 2022
C. vulgaris Diluted paper pulp & aquaculture (0.6:0.4)	25±2	7	0.04% CO_2 DNP	85 µmol photon/m².s⁻¹ Light:dark 24:0 hrs	1.31	0.12*	Daneshvar et al. 2018
Thalassiosira pseudonana Sterilized fishery effluent	28	7	Air @ 1 vvm	80 µmol photon/m².s⁻¹ Light:dark 12:12 hrs	0.11	NG	Wang et al. 2021
Isochrysis galbana Sterilized fishery effluent	28	7	Air @ 1 vvm	80 µmol photon/m².s⁻¹ Light:dark 12:12 hrs	0.14	NG	Wang et al. 2021
Monoraphidium sp. Dek19 & *Chlorella* sp. Urban (autoclaved final wastewater)	25 (10-28)	~9	Air @ 1.0-2.5 vvm	45-60 µmol photon/m².s⁻¹	NG	NG	Kirchener et al. 2022

Table 4. (Continued)

Microalgal species & wastewater	Culture conditions				Biomass yield (g/L)	Lipid yield (g/L)	Reference
	Temp. (°C)	Time (days)	Aeration	Light			
Coelastrella sp. Raw piggery effluent	25	4	5% CO_2 @ 0.8 vvm	Stage 1 Light:dark 0:48 hrs Stage 2. Light dark: NG @ 1250 μmol photon/m^2.s^{-1}	7.9	2.92-3.16*	Lee et al. 2021
C. sorokiniana Sterilized cooking cocoon effluent	25±2	7	Air @ 3.34 vvm	150 μmol photon/m^2.s^{-1} Light dark: 16:8	0.343	0.10*	Yang et al. 2022

DNP = details not provided NG = not given Vvm = volume medium per minute W/m^2 =
*calculated from data provided.

Algal specie	Wastewater	Culture conditions	Biomass	Lipid	Reference
				lipid content: 42% biodiesel yield: 3.5 g/L direct TE with acid catalyst and methanol and hexane	(Nambukrishnan and Singaram, 2022)

A comparison of the culture conditions and biomass and lipid productivities of some microalgal species grown in wastewater as a feedstock is provided in Table 4. As alluded to in Section 2, one of the biggest infield challenges for large scale oleaginous microalgal cultivation using wastewater as a feedstock is to maintain pure/axenic cultures. For 'real world' situations, MAB systems are therefore preferred (Section 2.3). More recently, it has been shown that co-culture of microalgae with fungi (including oleaginous yeasts), actinobacteria and cyanobacteria are also promising solutions (Cinq-Mars et al., 2022; Padri et al., 2022; Ling et al., 2014). As examples: (i) a microalgal-cyanobacterial system was able to effectively remove organics and N from urban wastewater with simultaneous production of lecanoric acid and ω-3 and -6 fatty acids (Cinq-Mars et al., 2022), (ii) a lipid yield of 25% was achieved by co-culturing the actinomycete, *Streptomyces thermocarboxydus* with the microalgae *C. sorokiniana* in cassava wastewater (Padri et al., 2022), and (iii) respective removal rates of 95%, 51% and 89% for COD, N and P and a lipid yield of 4.6 g L^{-1} was achieved by co-culture of the oleaginous yeast *Rhodosporidium toruloides* with the microalgae *C. pyrenoidosa* in mixed distillery and urban effluent (Ling et al., 2014). In large scale non-sterile environments like those encountered in MAB systems there is still a concern that undesirable microbial species may become dominant. This can be countered by constant inoculation with selected oleaginous species suited and acclimated to growth in a particular type of wastewater using hybrid systems (Section 3.2.3). Monitoring of the microbial diversity can assist with troubleshooting (Cinq-Mars et al., 2022).

Wherever possible, it is important to optimize the photoperiod (day: light) to enhance the growth and metabolism of the desired heterotrophic, autotrophic and/or mixotrophic oleaginous microalgae (Matos et al., 2017). The biomass productivity and the rate of lipidogenesis are also affected by the balance of macronutrients, particularly the C/N ratio. Provided they can be found in close proximity to one another, more favourable nutrient ratios may be achieved by mixing different wastewaters, for example by diluting high ammonia poultry wastewater with lower ammonia effluent. This strategy has already shown promise for increasing the biomass productivity of *C. sorokiniana* (Cui et al., 2020).

Amongst other elements, Fe is a key micronutrient for microalgal lipid synthesis and storage, and acetate is a key substrate for fatty acid synthesis. Iron is a co-factor in functional enzymes such as cytochrome oxidases and Fe/S proteins (Liu et al., 2021a). Supplementation of wastewater with key micronutrients and/or substrates such as Fe and/or acetate can therefore increase the rate of lipid synthesis and accumulation, as demonstrated by (i) the addition of 2.14×10^{-4} mol L^{-1} of Fe^{2+} and 2.0 g L^{-1} of sodium acetate to urban wastewater, which improved the respective dry cell weight, lipid content and lipid productivity of *C. pyrenoidosa* by 185%, 248% and 671% (Lui et al., 2021a), and (ii) the addition of 1 g L^{-1} sodium acetate to urban wastewater, which increased removal of N and P and lipid productivity of *T. obliquus*, resulting in an ultimate lipid yield of 2.08 mg $L^{-1}.d^{-1}$ (Liu et al., 2021a). Other chemical inducers such as Ca and plant growth regulators like gamma amino butyric acid (GABA) act as important signaling molecules that are involved in regulating cellular metabolism, C/N balance and stress acclimation in microalgal cells. Salinity stress, especially in the presence of chemical inducers, can induce biomass and lipid productivity of microalgal cells as demonstrated when adding 2.5 g L^{-1} NaCl and 0.5 mM GABA or 1.0 mM $CaCl^2$ to cultures of *Ankistrodesmus* sp. (Zhao et al., 2022). In addition to nutrient balancing, light and temperature selection, and the addition of chemical inducers and stress generation, another key factor that can affect the overall productivity of ABWBs is the selection of the most suitable growth phase for cell harvesting (Corredor et al., Gao et al., 2019). Higher lipid productivities of *C. vulgaris* and *T. obliquus* grown in a membrane photobioreactor containing secondary urban effluent have been achieved when harvested in stationary phase (29.8% and 36.9%, respectively), than when harvested in logarithmic phase (16.3% and 20.2%, respectively (Gao et al., 2019). Therefore, consideration and selection of suitable mixotrophic co-cultures, physicochemical conditions, wastewater types, and optimal harvesting in terms of stage of growth are all required to maximize concurrent wastewater remediation efficiency, biomass and lipid yield in ABWBs.

Harvesting of Microalgal Biomass

It has been calculated that 20-30% of the operating costs of microalgal biorefineries are related to biomass harvesting/recovery which involves concentration of biomass by removing excess water (Mahata et al., 2021; Vasistha et al., 2021). Selection of efficient, economically feasible and environmentally sustainable processes for harvesting is one of the biggest challenges faced by ABWBs. Some of the key factors to consider are: (i) the characteristics of microalgal cells (size, shape/structure, specific gravity, and charge), (ii) the operational scale, (iii) the capital equipment and running costs, and (iv) the wastewater characteristics (Mata et al., 2010; Min et al., 2022). Depending upon the cellular size and density of algal cells, singular or combined harvesting technique/s are applied (Laamanen et al., 2021). In general, filamentous microalgae (e.g., *Cladophora* sp., *Oedogonium* sp., *Spirogyra* sp. *Tribonema* sp.) are easier to harvest than unicellular species and are subject to less predation in bioreactors (Huo et al., 2019).

Research efforts are currently being directed towards methods that are suitable for harvesting microalgal or microalgal/bacterial/fungal consortia from large scale HRAPs and/or PBRs (Rossi et al., 2021). In this section, the discussion focusses on the harvesting methods that show promise for future large-scale applications or have been widely researched and/or applied. Although attempts have been made to immobilize algal species, for example on alginate beads (Mohseni et al., 2021a), the technology has a number of challenges and is likely to be prohibitively expensive at industrial scale. It has therefore only been assessed at laboratory scale to date and is not discussed further in this chapter.

Sedimentation and Flotation

Many unicellular microalgal cells have negative charges on their surfaces and tend to remain in suspension, while most filamentous green algae tend to float (Ghazvini et al., 2022). In general, microalgal cells sink more easily when grown as part of microbial consortia, so that biomass sedimentation tends to be more efficient in MAB systems than in pure microalgal growth PBRs, for example (Singh and Mishra, 2021; Soroosh et al., 2022). Sedimentation involves gravitational settling of high-density, high molecular weight microalgal biomass, while with flotation the cells are encouraged to float on the surface of the liquid fraction rather than settle. Sedimentation is simple and cost effective, and the integrity of the microalgal cells is preserved because no external forces are applied. It is commonly used in laboratory-based research due to its simplicity. However, at scale, it can be time-consuming if the biomass does not settle readily. This can be overcome to various extents if flocculation is combined with downstream settling.

Flotation relies on the formation of gaseous bubbles that cause the microalgal cells to coalesce and then migrate to the surface by lowering their relative density in the bulk liquid. The floating biomass can easily be harvested by mechanical skimming. Flotation efficiency is affected by the size and shape of the cells and the quality and quantity of the bubbles (Ghazvini et al., 2022; Lal & Das, 2016; Theoneste et al., 2016). Large-scale harvesting of cultivated microalgal biomass using flotation is relatively rapid and ensures high yield to area ratios. It can be expedited by addition of cationic surfactants like cetyltrimethylammonium bromide (CTAB) or biosurfactants such as rhamnolipid and bovine serum albumin (BSA). The surfactants increase the overall bubble surface area by promoting the formation of microbubbles *viz.* foam (Krishnan et al., 2022). Surfactant-assisted foam flotation tends to reverse the bubble surface charge. This permits hydrophobic electrostatic interaction between negatively charged algal cells with positively charged bubbles (Nie et al., 2022). The strong oxidant, ozone, can be used instead of surfactant addition to generate foam. The mechanism in this case is the release of amphiphilic cellular surfactant molecules from the microalgal cells own membranes via ozonolysis (Hernández et al., 2022). Ozone-air flotation has shown promise at laboratory scale, with more than 75% of *T. obliquus* biomass being harvested by this method (Hernández et al., 2022).

Centrifugation

Centrifugation works on the principle that particles/cells/solutions are separated according to density using centrifugal force (Khan et al., 2022). In conjunction with sedimentation, microalgae can be concentrated up to 99 times (de Souza Leite & Daniel, 2020). For lab-scale studies, bench top centrifuges are commonly used while for commercial scale applications examples include disc stack centrifuges, decanter centrifuges, tubular centrifuges and hydrocyclones. Despite the benefits of centrifugation, algal cells may be damaged, and scale-up is limited due to high capital and operational (energy) costs (Najjar & Abu-Shamleh, 2020).

Membrane Filtration

Microalgae with diverse morphologies (single celled, multi-celled, spherical, oval, filamentous) and sizes (3-30 µm) can be easily separated from the wastewater using membrane separation technology. Polyvinylchloride (PVC), polyacrylonitrile (PAN), polyvinylidene fluoride (PVDF), polyether sulfone (PES), and polytetrafluorethylene (PTFE) are examples of membrane materials widely used for filtration of microalgal cells (Yang et al., 2022). Membranes are classified according to the size of the particles that they retain: macro (≥ 10 µm), micro (0.1 to 10 µm), ultra-filtration (0.02 to 0.2 µm) and reverse osmosis (<than 0.001 µm) (Ghazvini et al., 2022). The speed of filtration is a function of the membrane pore size as well as the driving force (pressure applied) and the characteristics of the wastewater. Almost complete retention of microalgal biomass can be achieved in a continuous process without addition of chemicals or disruption of microalgal cell structure. In addition, the effluent can be recycled (Ghazvini et al., 2022; Zhao et al., 2020; Zhao et al., 2021a, Zhao et al., 2021b). On the negative side, membrane fouling is commonly encountered due to low permeate flux and deposition of microalgal extracellular organic matter (EOM) (Yang et al., 2022) and the technology is costly due to high capital expenditure, maintenance and energy requirements. However, membrane filtration technology is constantly becoming more economically viable and may very well become the harvesting method of choice over time.

Flocculation

Microalgal cells secrete EOM that surrounds the cells and imparts surface properties that are critical for floc formation (Henderson et al., 2010; Roselet et al., 2017; Vandamme et al., 2012; Vu et al., 2021; Wang et al., 2022; Xing et al., 2021). Flocculation can occur spontaneously, or it can rely on the application of a physical force, chemical- or bio-flocculant to a solution to drive the microalgal particles out of suspension. It is typically accompanied by downstream gravity sedimentation. Flocculation processes for microalgal harvesting range from conventional industrial methods such as chemical coagulation to innovative laboratory scale concepts based on, for example, microalgal cell biology (bio-flocculation) and evolving tools (magnetic-flocculation with micro or nanoparticles) (Muhammad et al., 2021; Ahmad et al., 2011; Lal & Das, 2016; Mahata et al., 2021; Vandamme et al., 2013).

Physical Flocculation
Physical flocculation is an environmentally friendly recovery process that involves the application of physical forces such as ultrasonication, electric or magnetic fields to induce flocculation without the addition of chemicals. Information on ultrasonic flocculation of microalgae is limited, but electro-flocculation has been well researched as it is perceived to be more promising at scale (Li et al., 2020). During electro-flocculation, negatively charged microalgae gravitate towards metallic anodes (for example, Al, Mg, Zn, Cu, Fe anodes), where the aggregated biomass can be collected (Krishnamoorthy et al., 2021). Factors such as mixing, temperature, pH, salinity, and the presence of chlorine impact on overall process efficiency (Mubarak et al., 2019; Vasistha et al., 2021). For example, higher flocculation efficiencies have been found for *C. vulgaris* and *Phaeodactylum tricornutum* at pH 6 (90-95%) than at pH 4 (80%). Nevertheless, microalgal recovery rates of 99%, 90% and 64% have been reported using Al, Mg and Fe anodes, respectively (Li et al., 2020; Yin et al., 2020). In general, it has been observed that current densities and rapidity of flocculation are directly proportional to one another (Vandamme et al., 2013). Electro-flocculation with Mg anodes appears to have relatively low energy requirements and short settling times, as described in a study by Yin et al. (2020) where settling was achieved under 10 min with the application of 40 volts. Results such as these are promising because wide-scale application of electro-flocculation for harvesting of microalgal biomass has been restricted in the past by high energy requirements (Vandamme et al., 2013). Power consumption for most microalgal harvesting is reported to be 0.3 to 2.0 kWh/kg, with 95% harvesting efficiencies for species such as *C. vulgaris, Phaeodactylum tricornutum, Closterium* sp., *Pediastrum* sp., *Cryptomonas* sp., *Staurastrum* sp. and blue green algae such as *Coelosphaerium* sp. and *Aphanizomenon* sp. at current densities of 3 mA/cm^2 for 30 min (Mubarak et al., 2019; Vasistha et al., 2021).

A new avenue of research involves the application of magnetic micro or nano particles such as magnetite (Fe_2O_3) for the creation of magnetic fields to induce microalgal flocculation. The magnetic particles can be recycled, and the process does not destroy the microalgal cells (Schobesberger et al., 2021).

Micro and nano magnetic particles have been successfully exploited to harvest cells of *C. vulgaris* (Zhu et al., 2018), *Scenedesmus* sp. (Abo Markeb et al., 2019) and *Nannochloropsis* sp. at laboratory scale.

Auto-Flocculation

Auto-flocculation is achieved without adding physical, chemical or biological flocculants, although it does rely on pH adjustment using chemicals, and in some cases manipulating the wastewater composition, particularly the O_2, N and Ca concentrations (Laamanen et al., 2021). In alkaline environments, the microalgal cells are no longer negatively charged, which promotes settling. In addition, the EOM can act as a biopolymer that interacts with OH^- forming bridges between the cells, promoting formation of large flocs (Gonzalez-Torres et al., 2017). Of the filamentous green algae, only the genera *Scenedusmus* and *Botryococcus* auto-flocculate (Xia et al., 2020) (Matter et al., 2019; Min et al., 2022). Despite its name, the fact that auto-flocculation requires chemical pH adjustment makes it unsuitable for large-scale harvesting of microalgal biomass from wastewater (Min et al., 2022).

Chemical Flocculation

Chemical flocculation is an effective harvesting method that involves utilization of large quantities of chemicals, metal coagulants, or cationic polymers. Flocculants/coagulants can be of organic or inorganic origin. Inorganic forms include NaOH, $Al_2(SO_4)_3$, $FeCl_3$, $Al_2Cl(OH)_5$ and $Fe_2(SO4)_3$ (Khan et al., 2022). Release of metallic ions from these compounds neutralizes the negatively charged microalgal cells and promotes floc formation, with high biomass recovery rates via downstream physical settling (Li et al., 2020; Chen et al., 2011a; Koley et al., 2017; Min et al., 2022). Chemical coagulation is tried and tested, but some additives are expensive, and the biomass becomes contaminated with metal ions when common inorganic flocculants are applied.

Organic flocculants work on the same principle of as inorganic flocculants. Those that have been used for successful harvesting of both freshwater and marine algal biomass are cationic organic polymers, including chitosan, starch, cellulose, tannin-based coagulants (TBCs), and (synthetic) polyacrylamides (Chen et al., 2011b; Li et al., 2020; Nguyen et al., 2022). Organic flocculants of 'natural' origin are typically viewed as more environmentally friendly than their inorganic counterparts (Hernández et al., 2022; Li et al., 2020). There are many examples of studies that show the effectiveness of natural organic flocculants for microalgal harvesting, for example >99% harvesting efficiency of a native algal consortium containing

Scenedesmus sp., *Chlorella* sp., *Schroderia* sp., and *Chlamydomonas* sp. using 20 mg/L of low-molecular-weight chitosan from shrimp shells (Acosta-Ferreira et al., 2020) and >90% biomass recovery of robust flocs from microalgae with TBCs (Teixeira et al., 2022). Flocculation can be enhanced using combined methods, as demonstrated with chitosan and ultrasonication, but sophisticated set-ups such as these have not been applied at scale (Li et al., 2020). There are multiple challenges associated with the use of 'natural' coagulants, including sourcing, modifying, transporting and ultimately removing these substances from the harvested biomass. The modification processes are in many instances chemically based and/or energy intensive, which nullifies the touted 'natural' (sustainable) origin of the flocculants themselves. In addition, as with inorganic flocculants, the harvested biomass contains coagulants and/or coagulant residues that may interfere with transesterification.

In terms of synthetic organic coagulants like polyacrylamides, floc formation is largely dependent on the molecular weight and charge density of the EOM. High molecular weight EOM provides multiple binding sites for interaction between the microalgal cells and the flocculant, promoting floc formation (Cai et al., 2022). For cyanobacteria such as *Microcystis aeruginosa, Microcystis wesenbergii, Cylindrospermopsis raciborskii,* and *Phormidium ambiguum*, floc formation can be achieved with >100 kDa EOM fractions, while for green alga like *Chlorella* sp., 30–100 kDa EOM fractions are required. Currently, modified polyacrylamide chains and magnetic polyacrylamides are being developed to increase microalgal harvesting efficiency, process sustainability and flexibility, and flocculant biodegradability and recyclability (Shaikh et al., 2021).

Bio-Flocculation

Microbial EOM itself can act as a bio-flocculant, and some bacteria produce specific substances such as poly γ-glutamic acid (γ-PGA) that effectively promote flocculation of microalgal cells. Bio-flocculation can be achieved by addition of bio-flocculants produced *ex-situ*, or by co-culturing microalgal cells with flocculant-producing bacterial strains *in-situ*. Alternatively, co-culturing microalgal species with filamentous fungi promotes flocculation by entanglement of the microalgae within the filament matrices or co-pelleting of the microalgae with the fungi. The co-pelleting mechanism results in higher mechanical stability of the cells than entanglement (Chu et al., 2021a). Co-culturing studies have yielded promising results, for example, 95-98% recovery of *C. vulgaris* and *C. protothecoides* co-cultured with γ-PGA-producing strains of *Bacillus licheniformis* and *Bacillus subtilis* (Ndikubwimana et al., 2016), 98% recovery of *Chlorella* sp. co-cultured with filamentous *Penicillium* spp. (Min et al., 2022), and 99% recovery of *C. vulgaris* co-pelleted with *Aspergillus oryzae* (Chu et al., 2021b).

In terms of ABWBs, co-culturing may be a viable option going forward for harvesting of biomass with concurrent wastewater remediation and lipid production. Co-pelleting of *C. vulgaris* with *Aspergillus niger* has been shown to effectively remove nutrients from cassava effluent (Padri et al., 2022), for example, and co-cultures of *T. obliquus* with *Rhodotorula glutinis* showed enhanced lipid production (60-70%) (Min et al., 2022; Yen et al., 2015).

A third mechanism for promoting bio-flocculation is to supplement the microalgal growth medium with an organic polymer like glycine to promote formation of EOM as a bio-flocculant as described by a number of researchers with *Botryococcus braunii* (Okoro et al., 2019; Xu et al., 2014; Kim et al., 2013). However, this is not realistic for ABWBs. A more promising alternative is to thermally extract EOM from waste material such as anaerobic sludge *ex-situ* (Mahata et al., 2021).

Extraction of Microalgal Lipids

Microalgae can accumulate 100 times more oil per acre in comparison to oil-containing plant cells (Mubarak et al., 2015). Traditionally, for biodiesel production, lipids are chemically extracted from microalgae before transesterification (Figure 2). However, direct transesterification (combined extraction and transesterification) is more likely to succeed in helping to bridge the cost gap between microalgal biodiesel and plant-based biodiesel. Direct transesterification is discussed in detail in Section 3.6 and separate extraction procedures that are still widely used for research purposes are briefly outlined in this section.

Harvested microalgal biomass is usually dried before extraction. To save on the energy required for drying, research has been conducted on direct use of dewatered biomass instead. As water forms a barrier between lipids and non-polar organic solvents, polar and non-polar solvents are used to enhance the extraction efficiency (Ghasemi Naghdi et al., 2016). When compared with the conventional (Soxhlet and Bligh and Dyer) extractive methods with a simultaneous distillation and extraction process (SDEP) of dewatered biomass of *D. salina* and *N. oculata* using d-limonene, a-pinene and p-cymene as solvents, similar lipid yields were obtained, but the energy consumption was substantially reduced (from 8.84 kWh to 2.15 kWh) using SDEP (Dejoye Tanzi et al., 2013).

Expellers and presses used for mechanical extraction of oils from plant oil seeds are not suitable for extraction of lipids from microalgae (Rawat et al., 2013; Vasistha et al., 2021). Standard methods that have been used for decades are Soxhlet extraction using n-hexane, phase extraction using chloroform: methanol (Bligh and Dyer method), and more recently, supercritical fluid extraction (SFE) (Adam et al., 2012). In addition to n-hexane, chloroform and methanol, other solvents such as benzene, isopropanol, and petroleum ether have also been used. Not only are these solvents expensive and in many cases highly volatile, but they are also toxic. Alternative methods such as SFE using CO_2 are therefore preferred from a sustainability perspective (Mubarak et al., 2015). Other new solvent-free methods have also been developed with the assistance of ultrasonication, electroporation, and microwave irradiation (Kumar et al., 2015). Currently, there are drawbacks to each of these. For example, ultrasonication generates a low-quality product, and there are issues with standardization of microwave-assisted extraction. Some scientists are of the opinion that combined extraction processes may address some of the historic problems. Examples include mechanically assisted solvent extraction, microwave assisted enzymatic/ solvent extraction, and pre-treatment of microalgal cells by thermal or osmotic shock before solvent extraction (Ghasemi Naghdi et al., 2016; Rawat et al., 2013; Sati et al., 2019). These strategies are intended to reduce the volume of solvents and energy consumption (Kumar et al., 2015). For ABWBs to be economically and environmentally successful, these new methods, including *in-situ* transesterification, direct saponification-acidification, and SFE need to be adopted in a cost-effective manner (Nagappan et al., 2019).

Biodiesel Production via Direct Algal Biomass Transesterification

Transesterification of extracted viscous microalgal lipids to fatty acid methyl esters (FAME) and/or fatty acid ethyl esters (FAEE) and glycerol is conducted using the same methodologies used for biodiesel generation from vegetable oil feedstocks using a range of different alcohols and catalysts. Biodiesel is recovered by phase separation using a variety of non-polar solvents such as chloroform and/or hexane, which are evaporated and recycled into the process. Generally, microalgal biodiesel produced using this approach is not economically competitive with biodiesel made from edible seed oils. However, in the case of ABWBs, feedstock costs are lower and more sustainable than the use of food crops for biodiesel. To further reduce costs and decrease the environmental footprint of microalgal biodiesel, studies have confirmed the feasibility of producing FAME/FAEE via direct (or *in-situ*) microalgal biomass transesterification, where lipid extraction and transesterification takes place in a single step using wet biomass. Direct transesterification of wet microalgae with H_2SO_4 as a catalyst can achieve comparable results to the 2-step process with a range of microalgal species including *Isochrysis zhangjiangensis*, *Nannochloropsis* sp., *Tetraselmis subcordiformis*, *A. platensis*, *Dunaliel-la salina*, *Chlamydomonas reinhartii*, *Synechocystis* sp. 6803, and *C. pyrenoidosa* (Liu et al., 2015). The residual biomass after direct transesterification contains carbohydrates, proteins and unreacted lipids, and it has been proposed that it is a suitable feedstock for anaerobic digestion to obtain energy from biogas, and digestate as a source of nutrients (Salam et al., 2016). The use of wet biomass as feedstock eliminates the energy required for drying the biomass before extraction, and strategies have been found to successfully overcome reduced biodiesel yields previously associated with the presence of water during transesterification (Kim et al., 2017). However, although solvents are recyclable, the larger volumes of biomass and solvents require more heating energy for direct transesterification of wet rather than dry biomass (Kim et al., 2017).

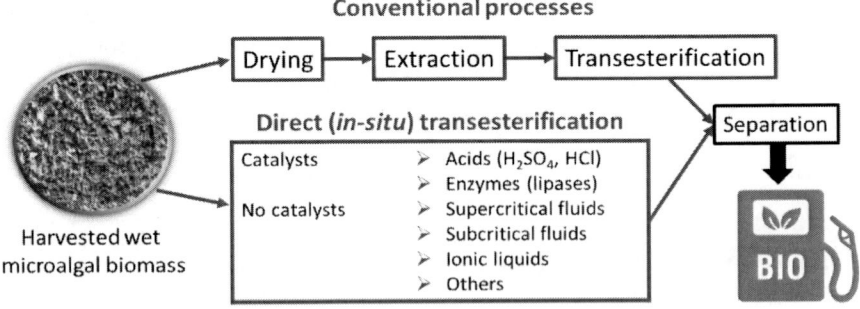

Figure 2. Schematic showing conventional versus direct transesterification of wet harvested algal biomass to biodiesel.

Acid Catalysts for Direct Transesterification

Some of the direct microalgal transesterification strategies are reliant of the addition of catalysts, while others are not. Basic catalysts are not suitable for substrates such as microalgal cells that contain high percentages of free fatty acids, particularly in the presence of water. Approaches based on 2-step extraction-transesterification using methanol and acid catalysts have been investigated. In the 1-step (direct) process, the wet microalgal biomass, methanol and acid catalysts (H_2SO_4 or HCl) are mixed and heated in 'one pot' under optimal conditions of temperature (90-150°C), wet cell weight, reaction time, and catalyst volume (Kim et al., 2015; Im et al., 2017; Kim et al., 2017) with achievable biodiesel yields of > 90% at lab-scale (Kim et al., 2015; Im et al., 2017). Heating lowers the required concentration of acid catalyst (Kim et al., 2015). Although H_2SO_4 is more widely used, it has been shown that higher yields can be achieved with HCl, and recovery of HCl requires less heating because of its lower boiling point. Chloroform can be used as a co-solvent with methanol and acid catalysts where it acts as and extractant and also reduces the concentrations of undesirable cations such as Na, Mg K and Ca that are typically present in high concentrations in microalgal biomass (Kim et al., 2015).

In a holistic study, direct transesterification of *C. pyrenoidosa* grown in 50 L outdoor HRAPs (raceways) containing paddy-soaked wastewater was able to convert 46.54% of algal lipids to FAME at optimal temperature (90°C) and ratios of methanol, H_2SO_4 and wet biomass (Umamaheswari et al., 2020). This study provided a proof of concept for ABWBs using direct transesterification with simultaneous removal rates of 75.89 ± 0.69% and 73.71 ± 0.75% for NH_3-N and PO_4^{2-}-P, respectively from the wastewater and lipids constituting 27.47 ± 1.41% of the biomass (Umamaheswari et al., 2020). While methanol is the most common alcohol substrate used in conjunction with acid catalysts for direct transesterification of microalgal biomass, comparable results have been found using ethanol in terms of FAME and FAEE yields (Lemões et al., 2016). Bioethanol, especially bioethanol fermented from waste or industry by-products is viewed as more sustainable from an environmental perspective. However, is more expensive in comparison to methanol which is typically derived through synthetic processes.

Lipases as Catalysts for Direct Transesterification

Lipases are perceived to be more environmentally friendly, albeit more expensive than acid catalysts for direct transesterification. Currently, the use of lipases as catalysts for biodiesel is not commercially viable because of the cost of the enzymes coupled with scale-up challenges. Nevertheless, studies have shown successful outcomes at lab-scale using both commercial and in-house enzymes. For example, the commercial enzyme Novozyme 435 (N435) from *Candida antarctica* was able to achieve FAME conversion of 99.5% from wet *Nannchloropsis gaditana* biomass under pressure (140 MPa), using t-butanol as a co-solvent with methanol in order to preserve enzyme activity and increase mass transfer rates (López et al., 2016). Successful direct transesterification of two metabolically engineered microalgal strains (*Chlorella emersonii* and *Pseudokirchneriella subcapitata*) has been achieved using in-house lipases from *Pseudomonas reinekei* and *Pseudomonas brenneri* with methanol and hexane as co-solvents (1:1). In this case, it was found that the FAME composition in the product was comparable to commercial biodiesel (Priyanka et al., 2020).

Catalyst-Free Methods for Direct Transesterification

Subcritical and Supercritical Fluids

In transesterification of wet microalgal biomass using subcritical and supercritical fluids, the water and methanol (or ethanol and/or co-solvent) obviates the need for catalysts, which is a major advantage in terms of material input and recycling costs. The use of microwave energy for heating theoretically has a lower energy demand than conventional heating and has the advantage that it concurrently disrupts the cell walls of microalgae promoting lipid extraction and increasing transesterification efficiency (Cheng et al., 2013). Although there are some impediments, microwave technology is constantly improving and industrial scale microwaves for continuous processing of biomass are already in existence (Priecel and Lopez, 2019).

The presence of free fatty acids (FFAs) does not have a negative effect on direct transesterification using supercritical fluids. The alcohol (methanol or ethanol) acts as a solvent, catalyst precursor and reactant, and the water from the wet biomass acts as a co-solvent (Patil et al., 2012; Najafabadi et al., 2015). Successful direct transesterification with supercritical fluids has been achieved with wet biomass from *Nannochloropsis* sp. (Patil et al., 2012) and *C. vulgaris* (Najafabadi et al., 2015), although the crude product may contain high concentrations of N (Patel and Hellgardt, 2016).

Although direct transesterification using supercritical fluids is a promising technology, significant energy is required to obtain the high temperatures and pressures required. This can be overcome to some extent by incorporating co-solvents with lower boiling points such as hexane, chloroform or diethyl ether in the reaction mix (Najafabadi et al., 2015). In general, subcritical fluid technology requires less energy than supercritical fluid technology, and shows promise at lab-scale, warranting further research as a prelude to introduction in the 'real world.' The principle of the reaction is that at the optimal temperature and input ratios, a homogenous methanol-water-oil phase is still formed, reducing the negative effect of water on transesterification as with supercritical fluid direct transesterification (Tsigie et al., 2012). Studies have shown good FAME yields (75-90%) from wet (80% wt%) *C. vulgaris* biomass at optimal temperatures (175-220°C) in methanol (Tsigie et al., 2012; Felix et al., 2019). Notably, a model based on lab-scale results found that the global warming potential (GWP) of the process inputs was significantly lower than the GWP from the conventional (drying-extraction-transesterification) approach (0.67 kg CO_2 g $FAME^{-1}$ vs 9.47 kg CO_2 g $FAME^{-1}$), with respective electrical energy demands of 5.994 MJ g $FAME^{-1}$ and 102.231 MJ g $FAME^{-1}$ (Felix et al., 2019).

Ionic Liquids
Ionic liquids are gaining traction as non-volatile, non-toxic, and thermostable replacements for conventional solvents. They are composed of inorganic and/or organic ions and can act as reagents, extractants, and catalysts. A recent study suggests that ionic liquids may have a role in ABWBS. Notwithstanding the fact that the research was conducted using dried biomass from *Nannochloropsis* sp., biodiesel yields obtained with 1-ethyl-3-methylimmidazolium methyl sulphate and methanol using microwave irradiation were higher than those obtained using a 2-step process, warranting further research into the place of Ionic liquids in ABWBs (Wahidin et al., 2018).

References

Abo Markeb, A., Llimós-Turet, J., Ferrer, I., Blánquez, P., Alonso, A., Sánchez, A., Moral-Vico, J., Font, X. 2019 The use of magnetic iron oxide based nanoparticles to improve microalgae harvesting in real wastewater. *Water Research* 159: 490–500 https://doi.org/10.1016/j.watres.2019.05.023.

Acosta-Ferreira, S., Castillo, O. S., Madera-Santana, J. T., Mendoza-García, D. A., Núñez-Colín, C. A., Grijalva-Verdugo, C., Villa-Lerma, A. G., Morales-Vargas, A. T., Rodríguez-Núñez, J. R. 2020 Production and physicochemical characterization of chitosan for the harvesting of wild microalgae consortia. *Biotechnology Reports* 28 https://doi.org/10.1016/j.btre.2020.e00554.

Adam, F., Abert-Vian, M., Peltier, G., Chemat, F. 2012 "Solvent-free" ultrasound-assisted extraction of lipids from fresh microalgae cells: A green, clean and scalable process.

Bioresource Technology 114: 457–465 https://doi.org/10.1016/j.biortech. 2012.02. 096.

Adesanya V. O., Cadena E., Scott S. A., Smith A. G. 2014 Life cycle assessment on microalgal biodiesel production using a hybrid cultivation system. *Bioresource Technology* 163: 343-355 http://dx.doi.org/10.1016/j.biortech.2014.04.051.

Ahmad, A. L., Mat Yasin, N. H., Derek, C. J. C., Lim, J. K. 2011 Optimization of microalgae coagulation process using chitosan. *Chemical Engineering Journal* 173: 879–882 https://doi.org/10.1016/j.cej.2011.07.070.

Almaguer M. A., Reyes Cruz Y., da Fonseca F. V. 2021 Combination of advanced oxidation process from microalgae aiming at recalcitrant wastewater treatment and algal biomass production: a review. *Environmental Processes* 8: 483-509 https://doi.org/10.1007/s40710-020-00492-x.

Aricila J. S., Céspedes D., Buitrón G. 2021 Influence of wavelength photoperiods and N/P ratio on wastewater treatment with microalgae-bacteria. *Water Science and Technology* 84: 712 https://doi.org/0.2166/wst.2021.257.

Avila R., Justo A., Carrero E., Crivillés E., Vicent T. Blánquez P. Water resource recovery coupling microalgae wastewater treatment and sludge co-digestion for bio-wastes valorisation at industrial pilot-scale. *Bioresource Technology* 343: 126080 https://doi.org/10.1016/j.biortech.2021.126080.

Bellucci M., Marazzi, F., Musatti A., Fornaroli R., Turolia A., Visigalli S., Bargna M., Bergna G., Canziani R., Mezzanote V., Rollini M., Ficara E. 2021 Assessment of annamox, microalgae and white-rot fungi-based processes for the treatment of textile wastewater. *PLOS ONE* 16: e0247452 https://doi.org/10.1371/journal.pone. 0247452.

Cai, Q., Song, K., Cai, P., Tian, C., Wang, C., Xiao, B. 2022 Harvesting of different microalgae through 100-μm-pore-sized screen filtration assisted by cationic polyacrylamide and specific extracellular organic matter. *Separation and Purification Technology* 280: 119918 https://doi.org/10.1016/j.seppur.2021.119918.

Casagli F., Rossi S., Steyer J. P., Bernard O., Ficara E. 2021 Balancing microalgae and nitrifiers for wastewater treatment: can inorganic carbon limitation cause an environmental threat? *Environmental Science and Technology* 55: 3840-3855 https://doi.org/10.1021/acs.est.0c05264.

Cheirsilp B., Thawechai T., Prasertsan P. 2017 Immobilized oleaginous microalgae for production of lipid and phytoremediation of secondary effluent from palm oil mill in fluidized bed photobioreactor. *Bioresource Technology* 241: 787–794 https://doi.org/10.1016/j.biortech.2017.06.016.

Chen, C. Y., Liu, C. H., Lo, Y. C., Chang, J. S. 2011a Perspectives on cultivation strategies and photobioreactor designs for photo-fermentative hydrogen production. *Bioresource Technology* 102: 8484–8492 https://doi.org/10.1016/j.biortech. 2011.05. 082.

Chen, C. Y., Yeh, K. L., Aisyah, R., Lee, D. J., Chang, J. S. 2011b Cultivation, photobioreactor design and harvesting of microalgae for biodiesel production: A critical review. *Bioresource Technology* 102: 71–81 https://doi.org/10.1016/j.biortech.2010.06.159.

Cheng J., Yu T., Li T., Zhou J., Cen K. 2013 Using wet microalgae for direct biodiesel production via microwave irradiation. *Bioresource Technology* 131: 531-535 https://doi.org/10.1016/j.biortech.2013.01.045.

Chu, R., Li, S., Yin, Z., Hu, D., Zhang, L., Xiang, M., Zhu, L. 2021b A fungal immobilization technique for efficient harvesting of oleaginous microalgae: Key parameter optimization, mechanism exploration and spent medium recycling. *Science of the Total Environment* 790: 148174 https://doi.org/10.1016/j.scitotenv.2021.148174.

Chu, R., Li, S., Zhu, L., Yin, Z., Hu, D., Liu, C., Mo, F. 2021a A review on co-cultivation of microalgae with filamentous fungi: Efficient harvesting, wastewater treatment and biofuel production. *Renewable and Sustainable Energy Reviews* 139: 110689 https://doi.org/10.1016/j.rser.2020.110689.

Cinq-Mars M., Bourdeau N., Marchand P., Desgagné-Penix I., Barnabé S. 2022 Characterization of two microalgae consortia grown in industrial wastewater for biomass valorization. *Algal Research* 62: 102628 https://doi.org/10.1016/j.algal.2021.102628.

Corredor L., Barnhart E. P., Parker A. E., Gerlach R., Fields M. W. 2021 Effect of temperature, nitrate concentration, pH and bicarbonate addition on biomass and lipid accumulation in the sporulating green alga PW95. *Algal Research* 53: 102148 https://doi.org/10.1016/j.algal.2020.102148.

Cui H., Ma H., Chen S., Yu J., Xu W., Zhu X., Gujar A., Ji C., Xue J., Zhang C., Li R. 2020. Mitigating excessive ammonia nitrogen in chicken farm flushing wastewater by mixing strategy for nutrient removal and lipid accumulation in the green alga *Chlorella sorokiniana*. *Bioresource Technology* 303(January): 122940 https://doi.org/10.1016/j.biortech.2020.122940.

Daneshvar E., Antikainen L., Koutra E., Kornaros M., Bhatnagar A. 2018. Investigation on the feasibility of Chlorella vulgaris cultivation in a mixture of pulp and aquaculture effluents: Treatment of wastewater and lipid extraction. *Bioresource Technology* 255: 104–110. https://doi.org/10.1016/j.biortech.2018.01.101.

Daneshvar E., Zarrinmehr M. J., Koutra E., Kornaros M., Farhadian O., Bhatnagar A. 2019 Sequential cultivation of microalgae in raw and recycled dairy wastewater: microalgal growth, wastewater treatment and biochemical oxidation. *Bioresource Technology* 273: 556-564 https://doi.org/10.1016/j.biortech.2018.11.059.

Dejoye Tanzi, C., Abert Vian, M., Chemat, F. 2013 New procedure for extraction of algal lipids from wet biomass: A green clean and scalable process. *Bioresource Technology* 134: 271–275 https://doi.org/10.1016/j.biortech.2013.01.168.

De Sousa Leite, L., Hoffman M. T., Daniel L. A. 2019 Microalgae cultivation for municipal and piggery wastewater treatment in Brazil. *Journal of Water Process Engineering* 31: 100821 https://doi.org/10.1016/j.jwpe.2019.100821.

de Souza Leite, L., Daniel, L. A. 2020 Optimization of microalgae harvesting by sedimentation induced by high pH. *Water Science and Technology* 82: 1227–1236 https://doi.org/10.2166/wst.2020.106.

Di Caprio F., Altimari P., Iaquaniello G., Toro L., Pagnanelli F. 2019 Heterotrophic cultivation of T. obliquus under non-axenic conditions by uncoupled supply of nitrogen and glucose. *Biochemical Engineering Journal* 145: 127-136 https://doi.org/10.1016/j.bej.2019.02.020.

Di Caprio F., Nguemna L. T., Stoller M. Giona M., Pagnanelli F. 2021 Microalgae cultivation by uncoupled nutrient supply in sequencing batch reactor (SBR) with olive mill wastewater treatment. *Chemical Engineering Journal* 410: 128417 https://doi.org/10.1016/j.cej.2021.128417.

Ding Y., Guo Z., Mei J., Liang Z., Li Z., Hou X. 2020 Investigation into the novel microalgae membrane bioreactor with internal circulating fluidized bed for marine aquaculture wastewater treatment. *Membranes* 10: 353 https://doi:10.3390/membranes10110353.

Ding Y., Wang S., Ma H., Ma B., Guo Z., You H., Mei J., Hou X., Liang Z., Li Z. 2021 Effect of different influent conditions on biomass production and nutrient removal by aeration microalgae membrane bioreactor (ACFB-MMBR) system for mariculture wastewater treatment. *Membranes* 11: 874 https://doi.org/10.3390/ membranes11110874.

Do J-M., Jo S-W., Kim I-S., Na H., Lee J. H., Kim H. S., Yoon H-S. 2019 A feasibility study of wastewater treatment using domestic microalgae and analysis of biomass for potential applications. *Water* 11: 2294 https://doi.org/10.3390/w11112294.

El-Sheekh M., El-Dalatony M. M., Thakur N., Zheng Y., El-Sayed S. 2021 Role of microalgae and cyanobacteria in wastewater treatment: genetic engineering and omics approaches. *International Journal of Environmental Science and Technology* https://doi.org/10.1007/s13762-021-03270-w.

Felix C., Ubando A., Madrazo C., Gue I. H., Sutanto S., Tran-Nguyen P. L., Go A. W., Ju Y. H., Culaba A., Chang J. S., Chen W-H. 2019 Non-catalytic in-situ (trans) esterification of lipids in wet microalgae Chlorella vulgaris under subcritical conditions for the synthesis of fatty acid methyl esters. *Applied Energy* 248: 526-537 https://doi.org/10.1016/j.apenergy.2019.04.149.

Feng X., Chen Y., Lv J., Han S., Tu R., Xu Z., Jin W., Ren N. Enhanced lipid production by *Chlorella pyrenoidosa* through magnetic field pretreatment of wastewater and treatment of microalgae-wastewater culture solution: Magnetic field treatment modes and conditions. *Bioresource Technology* 306: 123102 https://doi.org/10.1016/j.biortech.2020.123102.

Flores-Salgado G., Thalasso F., Buitrón G., Vital-Jácome M. 2021 Kinetic Characterization of microalgal-bacterial systems: Contributions of microalgae and heterotrophic bacteria to the oxygen balance in wastewater treatment. *Biochemical Engineering Journal* 165: 107819 https://doi.org/10.1016/j.bej.2020.1078.

Ganeshkumar V., Subashchandrabose S. R., Dharmarajan R., Venkateswarlu K., Naidu R., Megharaj M. 2018 Use of mixed wastewaters from piggery and winery for nutrient removal and lipid production by *Chlorella* sp. MM3. Bioresource Technology 256: 254–258. https://doi.org/10.1016/j.biortech.2018.02.025.

Gao B., Liu J., Zhang C., Van de Waal D. B. 2018 Biological stoichiometry of oleaginous microalgal lipid synthesis: The role of N:P supply ratios and growth rate on microalgal elemental and biochemical composition. *Algal Research* 32: 353–361 https://doi.org/10.1016/j.algal.2018.04.019.

Gao F., Cui W., Xu J. P., Li C., Jin W. H., Yang H. L. 2019 Lipid accumulation properties of Chlorella vulgaris and *Scenedesmus obliquus* in membrane photobioreactor (MPBR) fed with secondary effluent from municipal wastewater treatment plant. *Renewable Energy* 136: 671–676 https://doi.org/10.1016/j.renene.2019.01.038.

García D., de Godos I., Domínguez C., Turial S., Bolado S., Muñoz R. 2019 A systematic comparison of the potential on microalgae-bacteria and purple phototrophic bacteria consortia for the treatment of piggery wastewater. *Bioresource Technology* 276: 18-27 https://doi.org/10.1016/j.biortech.2018.12.095.

García D., Posadas E., Grajeda C., Blanco S., Martínez-Páramo S., Acién G., García-Encina P., Bolado S., Muñoz R. 2017 Comparative evaluation of piggery wastewater treatment in algal-bacterial photobioreactors under indoor and outdoor conditions. *Bioresource Technology* 245: 483-490 http://dx.doi.org/10.1016/j.biortech.2017.08.135.

García-Galán M. J., Arashiro L., Santos L. H. M. L. M., Insa S., Rodríguez-Mozaz S., Barceló D., Ferrer I., Garff M. 2020 Fate of priority pharmaceuticals and their main metabolites and transformation products in microalgae-based wastewater treatment systems. *Journal of Hazardous Materials* 390: 121771 https://doi.org/10.1016/j.jhazmat.2019.121771.

Geng Y., Cui D., Yang L., Xiong Z., Pavlostathis S. G., Shao P., Zhang Y., Luo X., Luo S. 2022 Resourceful treatment of harsh high-nitrogen rare earth element tailing (REEs) wastewater by carbonate activated *Chlorococcum* sp. microalgae. *Journal of Hazardous Materials* 423: 127000 https://doi.org/10.1016/j.jhazmat.2021.127000.

Ghasemi Naghdi, F., González González, L. M., Chan, W., Schenk, P. M. 2016 Progress on lipid extraction from wet algal biomass for biodiesel production. *Microbial Biotechnology* 9: 718–726 https://doi.org/10.1111/1751-7915.12360.

Ghazvini, M., Kavosi, M., Sharma, R., Kim, M. 2022 A review on mechanical-based microalgae harvesting methods for biofuel production. *Biomass and Bioenergy* 158: 106348 https://doi.org/10.1016/j.biombioe.2022.106348.

González-Camejo J., Aparicio S., Pachés N., Borrás L., Seco A. 2022 Comprehensive assessment of the microalgae-nitrifying bacteria competition in microalgae-based wastewater treatment systems: Relevant factors, evaluation methods and control strategies. *Algal Research* 61: 102563 https://doi.org/10.1016/j.algal.2021.102563.

Gonzalez-Torres, A., Rich, A. M., Marjo, C. E., Henderson, R. K. 2017 Evaluation of biochemical algal floc properties using Reflectance Fourier-Transform Infrared Imaging. *Algal Research* 27: 345–355 https://doi.org/10.1016/j.algal.2017.09.017.

Gupta S., Pawar S. B., Pandey R. A. 2019 Current practices and challenges in using microalgae for treatment of nutrient rich wastewater from agro-based industries. *Science of the Total Environment* 687: 1107-1126 https://doi.org/10.1016/j.scitotenv.2019.06.115.

Gutierrez J., Kwan T. A., Zimmerman J. B., Peccia J. 2016 Ammonia inhibition in oleaginous microalgae. *Algal Research* 19: 123–127 https://doi.org/10.1016/j.algal.2016.07.016.

Han W., Jin W., Li Z., Wei Y., He Z., Chen C., Qin C., Chen Y., Tu R., Zhou X. 2021 Cultivation of microalgae for lipid production using municipal wastewater. *Process*

Safety and Environmental Protection 155: 155–165 https://doi.org/10.1016/j.psep. 2021.09.014.

Hariz H. B., Takriff M. S., Yasin N. H. M., Ba-Abbad M. M., Hakimi N. I. N. M. 2019 Potential of the microalgae-based integrated wastewater treatment and CO_2 fixation system to treat Palm Oil Mill Effluent (POME) by indigenous microalgae: *Scenedesmus* sp. and *Chlorella* sp. *Journal of Water Process Engineering* 32: 100907 https://doi.org/10.1016/j.jwpe.2019.100907.

Heidari M., Kariminia H-R., Shayegan J. 2016 Effect of culture age and initial inoculum size on lipid accumulation and productivity in a hybrid cultivation system of Chlorella vulgaris. *Process Safety and Environmental Protection* 104: 111-122 http://dx.doi.org/10.1016/j.psep.2016.07.012.

Henderson, R. K., Parsons, S. A., Jefferson, B. 2010 The impact of differing cell and algogenic organic matter (AOM) characteristics on the coagulation and flotation of algae. *Water Research* 44: 3617–3624 https://doi.org/10.1016/j.watres.2010.04.016.

Hernández, E. V., Monje-Ramírez, I., Velásquez-Orta, S. B., Gracia-Fadrique, J., Orta Ledesma, M. T. 2022 Surface activity of biomolecules released from microalgae harvested by ozone-flotation. *Environmental Technology & Innovation* 26: 102354 https://doi.org/10.1016/j.eti.2022.102354.

Hilares R. T., Bustos K. A., Sanchez Vera F. P., Colina Andrade G. J. Acid precipitation followed by microalgae (Chlorella vulgaris) cultivation as a new approach for poultry slaughterhouse wastewater treatment. *Bioresource Technology* 335: 125284 https://doi.org/10.1016/j.biortech.2021.125284.

Hu X., Meneses Y. E., Hassan A. A. Integration of sodium hypochlorite pretreatment with co-immobilized microalgae/bacteria treatment of meat processing wasteweater. *Bioresource Technology* 304: 122953 https://doi.org/10.1016/j.biortech.2020.122953.

Hu X., Meneses Y. E., Stratton J., Lau S. K., Subbiah J. 2021 Integration of ozone with co-immobilised microalgae-activated sludge bacterial symbiosis for efficient on-site treatment of meat processing wastewater. *Journal of Environmental Management* 285: 112152 https://doi.org/10.1016/j.jenvman.2021.112152.

Huo S., Chen J., Zhu F., Zou B., Chen X., Bahseer S., Cui F., Qian J. 2019 Filamentous microalgae Tribonema sp. Cultivation in the anerobic/oxic effluents of petrochemical wastewater for evaluating the efficiency of recycling and treatment. *Biochemical Engineering Journal* 145: 27-32 https://doi.org/10.1016/j.bej.2019.02.011.

Im H., Lee H., Park M. S., Yang J. W., Lee J. W. 2014 Concurrent extraction and reaction for the production of biodiesel from wet microalgae. *Bioresource Technology* 152: 534-537 https://doi.org/10.1016/j.biortech.

Kadir W. N. A., Lam M. K., Uemura Y. Lim J. W. 2018 Harvesting and pre-treatment of microalgae cultivated in wastewater for biodiesel production: a review. *Energy Conservation and Management* 171: 1416–1429 https://doi.org/10.1016/j.enconman.2018.06.074.

Khalaji M., Hosseini S. A., Ghorbani R., Rezaei H., Kronaros M., Koutra E. 2021 Treatment of dairy wastewater by microalgae for biofuels production. *Biomass Conversion and Biorefinery* https://doi.org/10.1007/s13399-021-01287-2.

Khan, S., Naushad, M., Iqbal, J., Bathula, C., Sharma, G. 2022 Production and harvesting of microalgae and an efficient operational approach to biofuel production for a sustainable environment. *Fuel* 311: 122543 https://doi.org/10.1016/j.fuel.2021.1225 43.

Kim B., Chang Y. K., Lee J. W. 2017 Efficient solvothermal wet in situ transesterification of Nannochloropis gaditana for biodiesel production. *Bioprocess and Biosystems Engineering* 40: 723-730 https://doi.org/10.1007/s00449-017-1738-6.

Kim B., Im H., Lee J. W. 2015 In situ transesterification of highly wet microalgae using hydrochloric acid. *Bioresource Technology* 185: 421-425 https://doi.org/10.1016/j.biortech.2015.02.092.

Kim, D. G., Oh, H. M., Park, Y. H., Kim, H. S., Lee, H. G., Ahn, C. Y. 2013 Optimization of flocculation conditions for *Botryococcus braunii* using response surface methodology. *Journal of Applied Phycology* 25: 875–882 https://doi.org/10.1007/s10811-012-9948-4.

Kirchner N. J., Hage A., Gomez J., Grayburn W. S., Holbrook G. P. 2022 Photosynthesis, competition, and wastewater treatment characteristics of the microalga *Monoraphidium* sp. Dek19 at cool temperatures. *Algal Research* 62: 102624 https://doi.org/10.1016/j.algal.2021.102624.

Koley, S., Prasad, S., Bagchi, S. K., Mallick, N. 2017 Development of a harvesting technique for large-scale microalgal harvesting for biodiesel production. *RSC Advances* 7: 7227–7237 https://doi.org/10.1039/c6ra27286j.

Kotoula D., Iliopoulou A., Irakleous-Palaiologou E., Gatidou G., Aloupi M., Antonopoulou P., Fountoulakis M. S., Stasinakis A. S. 2020 Municipal wastewater treatment by combining in series microalgae *Chlorella sorkiniana* and macrophyte Lemna minor: Preliminary results. *Journal of Cleaner Production* 271: 122704 https://doi.org/10.1016/j.jclepro.2020.122704.

Krishnamoorthy, N., Unpaprom, Y., Ramaraj, R., Maniam, G. P., Govindan, N., Arunachalam, T., Paramasivan, B. 2021 Recent advances and future prospects of electrochemical processes for microalgae harvesting. *Journal of Environmental Chemical Engineering* 9: 105875 https://doi.org/10.1016/j.jece.2021.105875.

Krishnan, A., Devasya, R., Hu, Y., Bassi, A. 2022 Fundamental investigation of biosurfactants-assisted harvesting strategy for microalgae. *Biomass and Bioenergy* 158: 106364 https://doi.org/10.1016/j.biombioe.2022.106364.

Kulal D. K., DCosta C., Loni P. C., Some S., Kalambate P. K. 2020 Cyanobacteria: as a promising candidate for heavy-metals removal. In P. Singh, A. Kumar, S. VK, & A. Shrivastava (Eds.), *Advances in Cyanobacterial Biology* 291–300 Academic Press https://doi.org/https://doi.org/10.1016/C2018-0-05196-8.

Kumar, R. R., Rao, P. H., Arumugam, M. 2015 Lipid extraction methods from microalgae: A comprehensive review. *Frontiers in Energy Research* 3: 1–9 https://doi.org/10.3389/fenrg.2014.00061.

Laamanen, C. A., Desjardins, S. M., Senhorinho, G. N. A., Scott, J. A. 2021 Harvesting microalgae for health beneficial dietary supplements. *Algal Research* 54: 102189 https://doi.org/10.1016/j.algal.2021.102189.

Lal, A., Das, D. 2016 Biomass production and identification of suitable harvesting technique for *Chlorella* sp. MJ 11/11 and *Synechocystis* PCC 6803. *3 Biotech* 6: 1–10 https://doi.org/10.1007/s13205-015-0360-z.

Lavrinovič A., Murby F., Zīverte E., Mežule L., Juhna T. 2021 Increasing phosphorus uptake efficiency by phosphorus-starved microalgae for municipal wastewater posttreatment. *Microorganisms* 9: 1598 https://doi.org/10.3390/microorganisms9081598.

Lee S. A., Ko S. R., Lee N., Lee J. W., Le V. Van, Oh H. M., Ahn C. Y. 2021 Two-step microalgal (*Coelastrella* sp.) treatment of raw piggery wastewater resulting in higher lipid and triacylglycerol levels for possible production of higher-quality biodiesel. *Bioresource Technology* 332: 125081 https://doi.org/10.1016/j.biortech.2021.125081.

Lemões J. S., Alves Sobrinho R. C. M., Farais S. P., de Moura R. R., Primel E. G., Abreu P. C., Martins A. F., Montes D'Oca M. G. 2016 Sustainable production of biodiesel from microalgae by direct transesterification. *Sustainable Chemistry and Pharmacy* 3: 33-38 https://doi.org/10.1016/j.scp.2016.01.002.

Li, S., Hu, T., Xu, Y., Wang, J., Chu, R., Yin, Z., Mo, F., Zhu, L. 2020. A review on flocculation as an efficient method to harvest energy microalgae: Mechanisms, performances, influencing factors and perspectives. *Renewable and Sustainable Energy Reviews* 131: 110005 https://doi.org/10.1016/j.rser.2020.110 005.

Ling J., Nip S., Cheok W. L., de Toledo R. A., Shim H. 2014 Lipid production by a mixed culture of oleaginous yeast and microalga from distillery and domestic mixed wastewater. *Bioresource Technology* 173: 132–139 https://doi.org/10.1016/j.biortech.2014.09.047.

Ling Y., Sun L.ping, Wang S.ying, Lin C. S. K., Sun Z., Zhou Z. gang. 2019 Cultivation of oleaginous microalga *Scenedesmus obliquus* coupled with wastewater treatment for enhanced biomass and lipid production. *Biochemical Engineering Journal* 148: 162–169 https://doi.org/10.1016/j.bej.2019.05.012.

Liu J., Liu Y., Wang H., Xue S. 2015 Direct transesterification of fresh microalgal cells. *Bioresource Technology* 176: 284-287 https://doi.org/10.1016/j.biortech.2014.10.094.

Liu J, Song Y., Qiu W. 2017 Oleaginous microalgae *Nannochloropsis* as a new model for biofuel production: Review & analysis. *Renewable and Sustainable Energy Reviews* 72: 154–162 https://doi.org/10.1016/j.rser.2016.12.120.

Liu J., Yin J., Ge Y., Han H., Liu M., Gao F. 2021b Improved lipid productivity of *Scenedesmus obliquus* with high nutrient removal efficiency by mixotrophic cultivation in actual municipal wastewater. *Chemosphere* 285: 131475 https://doi.org/10.1016/j.chemosphere.2021.131475.

Liu J. Z., Yin J. Y., Han H. F., Ge Y. M., Wang Z. Y., Bao X. Y., Gao F. 2021a Enhancements of lipid productivity and phosphorus utilization efficiency of *Chlorella pyrenoidosa* by iron and acetate supplements in actual municipal wastewater. *Renewable Energy* 170: 927–935. https://doi.org/10.1016/j. renene.2021.01.148.

Liu X. L., Zhao G. P., Zhand H. K., Zhai Q. Y., Wang Q. 2022 Microalgae-based swine wastewater treatment: strain screening conditions optimization, physiological activity and biomass potential. *Science of the Total Environment* 807: 151008 https://doi.org/10.1016/j.scitotenv.2021.151008.

López E. N., Medina A. R., Cerdin L. E., Moreno P. A. G., Sánchez M. D. M., Grima E. M. 2016 Fatty acid methyl ester production from wet microalgal biomass by lipase-catalyzed transesterification. *Biomass and Bioenergy* 93: 6-12 https://doi.org/10.1016/j.biombioe.2016.06.018.

Mahata, C., Dhar, S., Ray, S., Das, D. 2021 Flocculation characteristics of anaerobic sludge driven-extracellular polymeric substance (EPS) extracted by different methods on microalgae harvesting for lipid utilization. *Biochemical Engineering Journal* 167 https://doi.org/10.1016/j.bej.2020.107898.

Makut B. B., Das D., Goswami G. 2019 Production of microbial biomass feedstock via co-cultivation of microalgae-bacteria consortium coupled with effective wastewater treatment: A sustainable approach. *Algal Research* 37: 228-239 https://doi.org/10.1016/j.algal.2018.11.020.

Mantovani M., Marazzi F., Fornaroli R., Bellucci M., Ficara E., Mezzanotte V. 2020 Outdoor pilot-scale raceway as a microalgae-bacteria sidestream treatment in a WWTP. *Science of the Total Environment* 710: 135583 https://doi.org/10.1016/j.scitotenv.2019.135583.

Marazzi F., Bellucci M., Fantasia T., Ficara E., Mexxanotte V. Interactions between microalgae and bacteria in the treatment of wastewater from milk whey processing. *Water* 12: 297 https://doi.org/10.3390/w12010297.

Marchão L., Fernandes J. R., Sampaio A., Peres J. A., Tavares P. B., Lucas M. S. 2021 Microalgae and immobilized TiO_2/UV-A LEDs as a sustainable alternative for winery wastewater treatment. *Water Research* 203: 117464 https://doi.org/10.1016/j.watres.2021.117464.

Martín L. A., Popovich C. A., Martinez A. M. Damiani M. C., Leonardi P. I. 2016 Oil assessment of Halamphora coffeaeformis diatom growing in a hybrid two-stage system for biodiesel production. *Renewable Energy* 92: 127-135 http://dx.doi.org/10.1016/j.renene.2016.01.078

Mata, T. M., Martins, A. A., Caetano, N. S. 2010 Microalgae for biodiesel production and other applications: A review. *Renewable and Sustainable Energy Reviews* 14: 217–232 https://doi.org/10.1016/j.rser.2009.07.020.

Matich E. K., Ghafari M., Camgoz E., Caliskan E., Pfeifer B. A., Haznedaroglu B. Z., Atilla-Gokcumen G. E. 2018 Time-series lipidomic analysis of the oleaginous green microalga species *Ettlia oleoabundans* under nutrient stress. *Biotechnology for Biofuels* 11: 1–15 https://doi.org/10.1186/s13068-018-1026-y.

Matos Â. P., Cavanholi M. G., Moecke E. H. S., Sant'Anna E. S. 2017 Effects of different photoperiod and trophic conditions on biomass, protein and lipid production by the marine alga *Nannochloropsis gaditana* at optimal concentration of desalination concentrate. *Bioresource Technology* 224: 490–497 https://doi.org/10.1016/j.biortech.2016.11.004.

Matter, I. A., Hoang Bui, V. K., Jung, M., Seo, J. Y., Kim, Y. E., Lee, Y. C., Oh, Y. K. 2019 Flocculation harvesting techniques for microalgae: A review. *Applied Sciences* 9 https://doi.org/10.3390/app9153069.

Min, K. H., Kim, D. H., Ki, M. R., Pack, S. P. 2022 Recent progress in flocculation, dewatering, and drying technologies for microalgae utilization: Scalable and low-cost harvesting process development. *Bioresource Technology* 344: 126404 https://doi.org/10.1016/j.biortech.2021.126404.

Mohseni A. Fan L., Roddick F. A. 2021b Impact of microalgae species and solution salinity on algal treatment of wastewater reverse osmosis concentrate. *Chemosphere* 285: 131 487 https://doi.org/10.1016/j.chemosphere.2021.131487.

Mohseni A., Kube M., Fan L., Roddick F. A. 2021a Treatment of wastewater reverse osmosis concentrate using alginate-immobilised microalgae: Integrated impact of solution conditions on algal bead performance. *Chemosphere* 276: 130028 https://doi.org/10.1016/j.chemosphere.2021.130028.

Moreno-García A., Neri-Torres E. E., Mena-Cervantes V. Y., Altamirano R. H., Pineda-Flores G., Luna-Sánchez R., García-Solares M., Vazquez-Arenas J., Suastes-Rivas J. K. 2021 Sustainable biorefinery associated with wastewater treatment of Cr (III) using a native microalgae consortium. *Fuel* 290: 119040 https://doi.org/10.1016/j.fuel.2020.119040.

Mubarak, M., Shaijaa, A., Suchithra, T. 2015 A review on the extraction of lipid from microalgae for biodiesel production. *Algal Research* 7: 117–123 https://doi.org/ https://doi.org/10.1016/j.algal.2014.10.008.

Mubarak, M., Shaijaa, A., Suchithra, T. 2015 A review on the extraction of lipid from microalgae for biodiesel production. *Algal Research* 7: 117–123 https://doi.org/ https://doi.org/10.1016/j.algal.2014.10.008.

Mubarak, M., Shaija, A., Suchithra, T. V. 2019 Flocculation: An effective way to harvest microalgae for biodiesel production. *Journal of Environmental Chemical Engineering* 7: 103221 https://doi.org/10.1016/j.jece.2019.103221.

Muhammad, G., Alam, M. A., Mofijur, M., Jahirul, M. I., Lv, Y., Xiong, W., Ong, H. C., Xu, J. 2021 Modern developmental aspects in the field of economical harvesting and biodiesel production from microalgae biomass. *Renewable and Sustainable Energy Reviews* 135: 110209 https://doi.org/10.1016/j.rser.2020.110209.

Nagappan, S., Devendran, S., Tsai, P. C., Dinakaran, S., Dahms, H. U., Ponnusamy, V. K. 2019 Passive cell disruption lipid extraction methods of microalgae for biofuel production – A review. *Fuel* 252: 699–709 https://doi.org/10.1016/j.fuel.2019.04.092.

Najafabadi H. A., Vossoughi M., Pazuki G. 2015 The role of co-solvents in improving the direct transesterification of wet microalgal biomass under supercritical condition. *Bioresource Technology* 193: 90-96 https://doi.org/10.1016/j.biortech.2015.06.045.

Najjar, Y. S. H., Abu-Shamleh, A. 2020 Harvesting of microalgae by centrifugation for biodiesel production: A review. *Algal Research* 51: 102046 https://doi.org/10.1016/j.algal.2020.102046.

Nambukrishnan V., Singaram J. 2022 Enhanced biodiesel production by optimizing growth conditions of *Chlorella marina* in tannery wastewater. *Fuel* 316: 123431 https://doi.org/10.1016/j.fuel.2022.123431.

Narala R. R. Garg S. G., Sharma K. K., Thomas-Hall S. R., Deme M., Li Y., Schenk P. M. 2016 Comparison of microalgae cultivation in photobioreactor, open raceway pond, and a two-stage hybrid system. *Frontiers in Energy Research* 4: 29 https://doi.org/10.3389/fenrg.2016.00029.

Ndikubwimana, T., Zeng, X., Murwanashyaka, T., Manirafasha, E., He, N., Shao, W., Lu, Y. 2016 Harvesting of freshwater microalgae with microbial bioflocculant: A pilot-scale study. *Biotechnology for Biofuels* 9: 1–11 https://doi.org/10.1186/s13068-016-0458-5.

Nguyen, L. N., Vu, H. P., Fu, Q., Abu Hasan Johir, M., Ibrahim, I., Mofijur, M., Labeeuw, L., Pernice, M., Ralph, P. J., Nghiem, L. D. 2022 Synthesis and evaluation of cationic polyacrylamide and polyacrylate flocculants for harvesting freshwater and marine microalgae. *Chemical Engineering Journal* 433: 133623 https://doi.org/10.1016/j.cej.2021.133623.

Nie, X., Zhang, H., Cheng, S., Mubashar, M., Xu, C., Li, Y., Tan, D., Zhang, X. 2022 Study on the cell-collector-bubble interfacial interactions during microalgae harvesting using foam flotation. *Science of the Total Environment* 806: 150901 https://doi.org/10.1016/j.scitotenv.2021.150901.

Oberholster P. J., Steyn M., Bothe A-M. 2021 A comparative study of improvement of phycoremediation using a consortium of microalgae in municipal wastewater treatment pond systems as an alternative solution to Africa's sanitation challenges. *Processes* 9: 1677 https://doi.org/10.3390/pr9091677.

Okoro, V., Azimov, U., Munoz, J., Hernandez, H. H., Phan, A. N. 2019 Microalgae cultivation and harvesting: Growth performance and use of flocculants - A review. *Renewable and Sustainable Energy Reviews* 115 https://doi.org/10.1016/j.rser.2019.109364.

Onyshchenko E., Blandin G., Comas J., Dvoretsky A. 2020 Influence of microalgae wastewater treatment culturing conditions of forward osmosis. *Environmental Science and Pollution Research* 27: 1234-1245 https://doi.org/10.1007/s11356-018-3607-5.

Paddock M. B., Fernández-Bayo J. D., van der Gheynst J. S. 2020 The effect of the microalgae-bacteria microbiome on wastewater treatment and biomass production. *Environmental Biotechnology* 104: 893-905 https://doi.org/10.1007/s00253-019-10246-x.

Padri, M., Boontian, N., Teaumroong, N., Piromyou, P., Piasai, C. 2022 Application of *Aspergillus niger* F5 as an alternative technique to harvest microalgae and as a phosphorous removal treatment for cassava biogas effluent wastewater. *Journal of Water Process Engineering* 46; 102524 https://doi.org/10.1016/j.jwpe.2021.102524.

Padri M., Boontian N., Teaumroong N., Piromyou P., Piasai C. 2022 Co-culture of microalga *Chlorella sorokiniana* with syntrophic *Streptomyces thermocarboxydus* in cassava wastewater for wastewater treatment and biodiesel production. *Bioresource Technology* 347: 126732 https://doi.org/10.1016/j.biortech.2022.126732.

Patel A., Karageorgou D., Rova E., Katapodis P., Rova U., Christakopoulos P., Matsakas L. 2020 An overview of potential oleaginous microorganisms and their role in biodiesel and omega-3 fatty acid-based industries. *Microorganisms* 8(3) https://doi.org/10.3390/microorganisms8030434.

Patel B., Hellgardt K. 2016 Hydrothermal liquefaction and in situ supercritical transesterification of algae paste. *RSC Advances* 6: 86560 https://doi.org/10.1039/c6ra11376a.

Patil P. D., Gude V. G., Mannarswamy A., Cooke P., Nirmalakhandan N., Lammers P., Deng S. 2012 Comparison of direct transesterification of algal biomass under supercritical methanol and microwave irradiation conditions. *Fuel* 97: 822-831 https://doi.org/10.1016/j.fuel.2012.02.037.

Plouviez M., Guiysse B. 2020 Nitrous oxide emission during microalgae-based wastewater treatment: current state of the art and implication for greenhouse gases budgeting.

Water Science and Technology 82.6: 1025-1030 https://doi.org/10.2166/wst.2020.3 04.

Pradana Y. S., Sudibyo H., Suyono E. A., Indarto, Budiman A. 2017 Oil algae extraction of selected microalgae species grown in monoculture and mixed cultures for biodiesel production. *Energy Procedia* 105: 277–282 https://doi.org/10.1016/j.egypro.2017.03.314.

Priecel P., Lopez-Sanchez J. A. 2019 Advantages and limitations of microwave reactors: from chemical synthesis to the catalytic valorization of biobased chemicals. *ACS Sustainable Chemistry and Engineering* 7: 3-21 https://doi.org/10.1021/acssuschemen g.8b03286.

Priyanka P., Kinsella G. K., Henehan G. T., Ryan B. J. 2020 Enzymatic in-situ transesterification of neutral lipids from simulated wastewater cultured *Chlorella emersonii* and *Pseudokirchneriella subcapitata* to sustainably produce fatty acid methyl esters. *Bioresource Technology Reports* 11: 100489 https://doi.org/10.1016/j.biteb.2020.100489.

Qiao S., Hou C., Zhou X. W. J. 2020 Minimizing greenhouse gas emission from wastewater treatment process by integrating activated sludge and microalgae processes. *Science of the Total Environment* 732: 139032 https://doi.org/10.1016/j.scitotenv.2020.139 032.

Qu W., Zhang C. Z., Chen X., Ho S-H 2021 New concept in swine wastewater treatment: development of a self-sustaining synergistic microalgae-bacteria symbiosis (ABS) system to achieve environmental sustainability. *Journal of Hazardous Materials* 418: 126264 https://doi.org/10.1016/j.jhazmat.2021.126264.

Raja S. W., Thunuja G. K., Karthikeyan S. Marimuthu S. 2022 Exploring the concurrent use of microalgae Coelastrella sp. for electricity generation and wastewater treatment. *Bioresource Technology Reports* 17: 100889 https://doi.org/10.1016/j.biteb.2021.100 889.

Rawat, I., Ranjith Kumar, R., Mutanda, T., Bux, F. 2013 Biodiesel from microalgae: A critical evaluation from laboratory to large scale production. *Applied Energy* 103: 444–467 https://doi.org/10.1016/j.apenergy.2012.10.004.

Razzak S. A., Hossain M. M., Lucky R. A., Bassi A. S., De Lasa H. 2013 Integrated CO2 capture, wastewater treatment and biofuel production by microalgae culturing - A review. *Renewable and Sustainable Energy Reviews* 27: 622–653 https://doi.org/ 10.1016/j.rser.2013.05.063.

Rearte T. A., Rodriguez N., Sabatté, Fabrizio de Iorio A. 2021 Unicellular microalgae vs. filamentous algae for wastewater nutrient recovery. *Algal Research* https://doi.org/ 10.1016/j.algal.2021.102442.

Ren Y., Deng J., Huang J., Wu Z., Yi L., Bi Y., Chen F. 2021 Using green alga *Haematococcus pluvialis* for astaxanthin and lipid co-production: Advances and outlook. *Bioresource Technology* 340: 125736. https://doi.org/10.1016/j.biortech.20 21.125736.

Roselet, F., Vandamme, D., Roselet, M., Muylaert, K., Abreu, P. C. 2017 Effects of pH, salinity, biomass concentration, and algal organic matter on flocculant efficiency of synthetic versus natural polymers for harvesting microalgae biomass. *Bioenergy Research* 10: 427–437 https://doi.org/10.1007/s12155-016-9806-3.

Rossi, S., Visigalli, S., Castillo Cascino, F., Mantovani, M., Mezzanotte, V., Parati, K., Canziani, R., Turolla, A., Ficara, E. 2021 Metal-based flocculation to harvest microalgae: a look beyond separation efficiency. *Science of the Total Environment* 799 https://doi.org/10.1016/j.scitotenv.2021.149395.

Salam K. A., Velasquez-Orta S. B., Harvey A. P. 2016 A sustainable integrated in situ transesterification of microalgae for biodiesel production and associated co-product – a review. *Renewable and Sustainable Energy Reviews* 65: 1179-1198 https://doi.org/10.1016/j.rser.2016.07.068.

Salih F. M. 2011 Microalgae tolerance to high concentrations of carbon dioxide: a review. *Journal of Environmental Protection* 02: 648–654 https://doi.org/10.4236/jep.2011.25074.

Sánchez-Zurano A., Lafarga T., del Mar Morales-Amaral M., Gómez-Serrano C., Fernández-Sevilla J. M., Acién-Fernández F. G., Molina-Grima E. 2021 Wastewater treatment using Scenedesmus almeriensis: effect of operational conditions on the composition of the microalgae-bacteria consortia. *Journal of Applied Phycology* 33: 3885-3897 https://doi.org/10.1007/s10811-021-02600-2.

Sati, H., Mitra, M., Mishra, S., Baredar, P. 2019 Microalgal lipid extraction strategies for biodiesel production: A review. *Algal Research* 38: 101413 https://doi.org/10.1016/j.algal.2019.101413.

Satpati G.G., Gorain P.C., Paul I., Pal R. 2016 An integrated salinity-driven workflow for rapid lipid enhancement in green microalgae for biodiesel application. *RSC Advances* 6: 112340-112355 https://doi.org/10.1039/c6ra23933a.

Schobesberger, M., Kopp Real, B., Meijer, D., Berensmeier, S., Fraga-García, P. 2021 Natural magnetite ore as a harvesting agent for saline microalgae *Microchloropsis salina*. *Bioresource Technology Reports* 15: 100798 https://doi.org/10.1016/j.biteb.2021.100798.

Shaikh, S. M. R., Hassan, M. K., Nasser, M. S., Sayadi, S., Ayesh, A. I., Vasagar, V. *2021* A comprehensive review on harvesting of microalgae using Polyacrylamide-Based Flocculants: Potentials and challenges. *Separation and Purification Technology* 277: 119508 https://doi.org/10.1016/j.seppur.2021.119508.

Siddiqui S., Suneetha Y. K. 2019 Effect of light intensity on algal growth in designed hybrid photobioreactor and biodiesel production. *International Journal and Advanced Technology* 9: 7497-7501. https://doi.org/10.35940/ijeat.A3126.109119.

Singh V., Mishra V. 2021 Exploring the effect of different combinations of predictor variable for the treatment of wastewater by microalgae and biomass production. *Biochemical Engineering Journal* 174: 108129 https://doi.org/10.1016/j.bej.2021.108129.

Spennati E., Casazza A. A., Converti A. 2020 Winery wastewater treatment by microalgae to produce low-cost biomass for energy purposes. *Energies* 13: 2490 https://doi.org/10.3390/en13102490.

Sushchik N. N., Kalacheva G. S., Zhila N. O., Gladyshev M. I., Volova T. G. 2003 A temperature dependence of the intra- and extracellular fatty-acid composition of green algae and cyanobacterium. *Russian Journal of Plant Physiology* 50: 374–380 https://doi.org/10.1023/A:1023830405898.

Sutherland D. L., Park J., Heubeck S., Ralph P. J., Craggs R. J. 2020 Size matters – Microalgae production and nutrient removal in wastewater treatment high rate algal ponds of three different sizes. *Algal Research* 45: 101734 https://doi.org/10.1016/j.algal.2019.101734.

Swain P., Tiwari A., Pandey A. 2020 Enhanced lipid production in Tetraselmis sp. by two stage process optimization using simulated dairy wastewater as feedstock. *Biomass and Bioenergy* 139: 105643 https://doi.org/10.1016/j.biombioe.2020.105643.

Teixeira, M. S., Speranza, L. G., da Silva, I. C., Moruzzi, R. B., Silva, G. H. R. 2022 Tannin-based coagulant for harvesting microalgae cultivated in wastewater: Efficiency, floc morphology and products characterization. *Science of the Total Environment* 807: 150776 https://doi.org/10.1016/j.scitotenv.2021.150776.

Theoneste, N., Jingyu, C., Zongyuan, X., Wenyao, S., Zeng, X., Ng, I.-S., Lu, Y. 2016 Flotation: A promising microalgae harvesting and dewatering technology for biofuels production. *Biotechnology Journal* 11 https://doi.org/https://doi.org/10.1002/biot.201500175.

Torres-Franco A., Passos F., Figueredo C., Mota C., Muñoz R. 2021 Current advances in microalgae-based treatment of high-strength wastewaters: challenges and opportunities to enhance wastewater treatment performance. *Review in Environmental Science and Biotechnology* 20: 209-235 https://doi.org/10.1007/s11157-020-09556-8.

Tsigie Y. A., Huynh L. H., Ismadji S., Engida A. M. Ju Y-H. 2012 *In situ* biodiesel production from wet Chlorella vulgaris under subcritical condition. *Chemical Engineering Journal* 123: 104-108 https://doi.org/10.1016/j.cej.2012.09.112.

Umamaheswari J., Kavitha M. S., Shanthakumar S. 2020 Outdoor cultivation of Chlorella pyrenoidsa in paddy-soaked wastewater and a feasibility study on biodiesel production from wet algal biomass through in-situ transesterification. *Biomass and Bioenergy* 143: 105853 https://doi.org/10.1016/j.biombioe.2020.105853.

Vandamme, D., Foubert, I., Fraeye, I., Muylaert, K. 2012 Influence of organic matter generated by chlorella vulgaris on five different modes of flocculation. *Bioresource Technology* 124: 508–511 https://doi.org/10.1016/j.biortech.2012.08.121.

Vandamme, D., Foubert, I., Muylaert, K. 2013 Flocculation as a low-cost method for harvesting microalgae for bulk biomass production. *Trends in Biotechnology* 31: 233–239 https://doi.org/10.1016/j.tibtech.2012.12.005.

Vasistha, S., Khanra, A., Clifford, M., Rai, M. P. 2021 Current advances in microalgae harvesting and lipid extraction processes for improved biodiesel production: A review. *Renewable and Sustainable Energy Reviews* 137: 110498 https://doi.org/10.1016/j.rser.2020.110498.

Vasistha, S., Khanra, A., Clifford, M., Rai, M. P. 2021 Current advances in microalgae harvesting and lipid extraction processes for improved biodiesel production: A review. *Renewable and Sustainable Energy Reviews* 137: 110498 https://doi.org/10.1016/j.rser.2020.110498.

Vu, H. P., Nguyen, L. N., Emmerton, B., Wang, Q., Ralph, P. J., Nghiem, L. D. 2021 Factors governing microalgae harvesting efficiency by flocculation using cationic polymers. *Bioresource Technology* 340: 125669 https://doi.org/10.1016/j.biortech.2021.125669.

Wahidin S., Idris A., Yusof N. M., Kamis N. H. H., Shaleh S. R. M. 2018 Optimization of the ionic liquid-microwave assisted one-step biodiesel production process from wet microalgal biomass. *Energy Conversion and Management* 171: 1397-1404 https://doi.org/10.1016/j.enconman.2018.06.083.

Walls L. E., Velasquez-Orta S. B., Romero-Frasca E., Leary P., Noguez Y. I., Ledesma M. T. A. 2019 Non-sterile heterotrophic cultivation of native wastewater yeast and microalgae for integrated municipal wastewater treatment and bioethanol production. *Biochemical Engineering Journal* 151: 108319 https://doi.org/10.1016/j.bej.2019.107319.

Wang H., Qi M., Bo Y., Zhou C., Yan X., Wang G., Cheng P. 2021 Treatment of fishery wastewater by co-culture of *Thalassiosira pseudonana* with *Isochrysis galbana* and evaluation of their active components. *Algal Research* 60: 102498 https://doi.org/10.1016/j.algal.2021.102498.

Wang M., Shi L-D., Lin D-X., Qui D-S., Chen J-P, Tao X-M., Tian G-M. 2020 Characteristics and performances of microalgal-bacterial consortia in a mixture of raw piggery digestate and anoxic aerated effluent. *Bioresource Technology* 309: 123363 https://doi.org/10.1016/j.biortech.2020.123363.

Wang, Q., Shen, Q., Wang, J., Zhao, J., Zhang, Z., Lei, Z., Yuan, T., Shimizu, K., Liu, Y., Lee, D. J. 2022 Insight into the rapid biogranulation for suspended single-cell microalgae harvesting in wastewater treatment systems: Focus on the role of extracellular polymeric substances. *Chemical Engineering Journal* 430: 132631 https://doi.org/10.1016/j.cej.2021.132631.

Watanabe M. W., Isdepsky A. 2021 Biocrude oil production by integrating microalgae polyculture and wastewater treatment: novel proposal on the use of deep water-depth polyculture of mixotrophic microalgae. *Energies* 14: 6992 https://doi.org/10.3390/en14216992.

Xia, Z., Li, J., Zhang, J., Zhang, X., Zheng, X., Zhang, J. 2020 Processing and valorization of cellulose, lignin and lignocellulose using ionic liquids. *Journal of Bioresources and Bioproducts* 5: 79–95 https://doi.org/10.1016/j.jobab.2020.04.001.

Xing, Y., Guo, L., Wang, Y., Zhao, Y., Jin, C., Gao, M., Ji, J., She, Z. 2021 An insight into the phosphorus distribution in extracellular and intracellular cell of Chlorella vulgaris under mixotrophic cultivation. *Algal Research* 60: 102482 https://doi.org/10.1016/j.algal.2021.102482.

Xu, X., Chen, C., Shen, Y. 2014 Flocculation of *Botryococcus braunii* with glycine. *Advanced Materials Research* 10: 877–880 https://doi.org/https://doi.org/10.4028/www.scientific.net/AMR.1004-1005.877.

Xu X. Q., Beardall J. 1997 Effect of salinity on fatty acid composition of a green microalga from an antarctic hypersaline lake. *Phytochemistry* 45: 655–658 https://doi.org/10.1016/S0031-9422(96)00868-0.

Yang, M., Xue, C., Li, L., Gao, Z., Liu, Q., Qian, P., Dong, J., Gao, K. 2022 Design and performance of a low-cost microalgae culturing system for growing *Chlorella sorokiniana* on cooking cocoon wastewater. *Algal Research* 62: 102607 https://doi.org/10.1016/j.algal.2021.102607.

Yang, Y., Zheng, M., Qiao, S., Zhou, J., Bi, Z., Quan, X. 2022 Electro-Fenton improving fouling mitigation and microalgae harvesting performance in a novel membrane photobioreactor. *Water Research* 210: 117955 https://doi.org/10.1016/j.watres.2021. 117955.

Ye, S., Gao L., Zhao J., An M., Wu H., Li M. 2020 Simultaneous wastewater treatment and lipid production by Scenedesmus sp. HXY2. *Bioresource Technology* 302: 122903 https://doi.org/10.1016/j.biortech.2020.122903.

Yen, H. W., Chen, P. W., Chen, L. J. 2015 The synergistic effects for the co-cultivation of oleaginous yeast-*Rhodotorula glutinis* and microalgae-*Scenedesmus obliquus* on the biomass and total lipids accumulation. *Bioresource Technology* 184: 148–152 https://doi.org/10.1016/j.biortech.2014.09.113.

Yin, Z., Zhu, L., Li, S., Hu, T., Chu, R., Mo, F., Hu, D., Liu, C., Li, B. 2020 A comprehensive review on cultivation and harvesting of microalgae for biodiesel production: Environmental pollution control and future directions. *Bioresource Technology* 301: 122804 https://doi.org/10.1016/j.biortech.2020.122804.

You X., Zhang Z., Guo L., Liao L., Liao Q., Wang Y., Zhao Y., Jin C., Gao M., She Z., Wang G. Integrating acidogenic fermentation and microalgae cultivation of bacterial-algal coupling system for mariculture wastewater treatment. *Bioresource Technology* 320: 124335 https://doi.org/10.1016/j.biortech.2020.124335.

Zhang A., Guo L., Liao Q., Gao M., Zhao Y., Jin C., She Z., Wang G. 2021 Bacterial-algal coupling system for high strength mariculture wastewater treatment: Effect of temperature on nutrient recovery and microalgae cultivation. *Bioresource Technology* 338: 125574 https://doi.org/10.1016/j.biortech.2021.125574.

Zhang C., Ren H. xue, Jiang L. 2021. Cultivation of *Chlorella protothecoides* in polyglutamic acid wastewater for cost-effective biodiesel production. *Arabian Journal of Chemistry* 14: 103108 https://doi.org/10.1016/j.arabjc.2021.103108.

Zhang Z., Ji H., Gong G., Zhang X., Tan, T. 2014 Synergistic effects of oleaginous yeast *Rhodotorula glutinis* and microalga *Chlorella vulgaris* for enhancement of biomass and lipid yields. *Bioresource Technology* 164: 93–99 https://doi.org/10.1016/j.biortech.2014.04.039.

Zhao Y., Qiao T., Gu D., Zhu L., Yu, X. 2022 Stimulating biolipid production from the novel alga *Ankistrodesmus* sp. by coupling salt stress and chemical induction. *Renewable Energy* 183: 480–490 https://doi.org/10.1016/j.renene.2021.11.034.

Zhao, Z., Li, Y., Muylaert, K., Vankelecom, I. F. J. 2020 Synergy between membrane filtration and flocculation for harvesting microalgae. *Separation and Purification Technology* 240: 116603 https://doi.org/10.1016/j.seppur.2020.116603.

Zhao, Z., Liu, B., Ilyas, A., Vanierschot, M., Muylaert, K., Vankelecom, I. F. J. 2021a Harvesting microalgae using vibrating, negatively charged, patterned polysulfone membranes. *Journal of Membrane Science* 618: 118617 https://doi.org/10.1016/j.memsci.2020.118617.

Zhao, Z., Muylaert, K., Vankelecom, I. F. J. 2021b Combining patterned membrane filtration and flocculation for economical microalgae harvesting. *Water Research* 198: 117181 https://doi.org/10.1016/j.watres.2021.117181.

Zhu, L., Li, Z., Hiltunen, E. 2018 Microalgae *Chlorella vulgaris* biomass harvesting by natural flocculant: Effects on biomass sedimentation, spent medium recycling and lipid extraction. *Biotechnology for Biofuels* 11: 1–10 https://doi.org/10.1186/s13068-018-1183-z.

Zkeri E., Iliopoulou A., Katsara A., Korda A., Aloupi M., Gatidou G., Fountoulakes M.S., Stasinakis A. S. 2021 Comparing the use of a two-stage MBBR system with a methanogenic MBBR coupled with a microalgae reactor for medium-strength dairy wastewater treatment. *Bioresource Technology* 323: 124629 https://doi.org/10.1016/j.biortech.2020.124629.

Chapter 2

Are Microreactors the Future of Biodiesel Synthesis?

Rosilene A. Welter[1,2], João L. Silva Jr.[3], Marcos R.P. de Sousa[1], Mariana G.M. Lopes[1], Osvaldir P. Taranto[1] and Harrson S. Santana[1,4,*]

[1]School of Chemical Engineering,
University of Campinas, Campinas, SP, Brazil
[2]College of Science and Engineering,
James Cook University,
Townsville, Queensland, Australia
[3]Federal University of ABC, Center for Engineering,
Modeling and Applied Social Sciences,
São Bernardo do Campo, SP, Brazil
[4]Center for Natural Sciences (CCN),
Federal University of São Carlos,
São Carlos, SP, Brazil

Abstract

Microfluidic devices or microdevices refer to systems with a characteristic length in the micrometer range. Systems in this size allow handling small quantities of reagents and samples, with reduced residence time, better control of chemical species concentration, high heat and mass transfers, and high surface/volume ratio. These characteristics led to the application of these microdevices in several areas, such as biological systems, energy, liquid-liquid extraction, food, agricultural sectors, pharmaceuticals, flow chemistry, microreactors, and

---

[*] Corresponding Author's Email: harrison.santana@gmail.com.

In: The Future of Biodiesel
Editor: Michael F. Simpson
ISBN: 979-8-88697-166-8
© 2022 Nova Science Publishers, Inc.

biodiesel synthesis. Microreactors are devices that have interconnected microchannels, in which small amounts of reagents are manipulated and react for a certain period of time. The traditional characteristics of microreactors are less quantities of reagents and samples, high surface area in relation to volume ($10000 \text{ m}^2 \text{ m}^{-3}$), reduction of resistance to heat and mass transfer, reduced reaction times, and narrower residence time distributions. In recent years, several studies have been carried out on biodiesel production in microreactors that explore the influence of operating conditions, mixing and reaction yield, numbering, and especially the microdevices design. Despite all the advantages of microreactors, the literature shows that there are only a few applications on an industrial scale. Two main reasons that hinder the adoption of this technology are the scale-up to a large enough volume to deliver the necessary production capacity and the costs related to industrial manufacturing microreactors. It is often stated that large-scale production of microreactors can be easily achieved by numbering-up. However, researches show that an incredibly high number of microdevices would be needed, which results in a technical unfeasibility and a strong impact on the construction costs of the industrial system. The present review aims to show whether microreactors can replace conventional biodiesel production processes and how this replacement technology could be carried out. The current chapter was divided into the following sections: Introduction, Synthesis and Purification of Biodiesel in Microreactors, Fundamentals of CFD, and Fundamentals of Scale-up. Finally, conclusions and future perspectives are exposed.

Keywords: biodiesel, microreactor, transesterification, Computational Fluid Dynamics

Introduction

The main energy source for power and transportation comes from fossil fuels. The increasing energy demand has resulted in the extensive reduction of oil reserves. In addition, the use of fossil fuels produces a large emission of CO_2, which exacerbates global warming and accentuates negative climate change. As an eco-friendly option, biodiesel has been considered a promising substitute for fossil fuel. Global biodiesel production increased from 6 billion liters per year in 2005 to 46 billion liters per year in 2020 (Ogunkunle and Ahmed 2019). It has advantages such as being four times easier to degrade (Demirbaş et al., 2008), emitting 86% fewer greenhouses gases (Voegele 2020), obtaining from renewable sources, such as crude vegetable oils, waste cooking oil (WCO), or animal fat (ASTM D6751–15, 2015). Biodiesel is a mixture of long-chain fatty acid alkyl ester (FAAE) produced by esterification and/or transesterification. Esterification occurs between free fatty acid and methanol in the presence of a catalyst, such as strong acid, having FAAE and water. Transesterification, the most common reaction, occurs between triglycerides and alcohol in the presence of a catalyst such as strong acid or alkali, producing FAAE and glycerol. The reactions usually happen on a macroscale by a continuous reactor or, more often, by a batch reactor; and take several hours to achieve high conversion requiring huge excess of alcohol, consuming extensive energy, and producing a large volume of residues (Bashir et al., 2022; Jachuck et al., 2009). These issues increase the process costs and hamper biodiesel's commercialization. Thus, improved processes are required to overcome these gaps. The microchemical plant has been shown as one of the most promising technologies.

A microchemical plant refers to a system with internal dimensions of 10-1000 μm (Balbino et al., 2016; Dimov et al., 2008; Santana et al., 2018a). According to Bannatham et al., (2021), the microscale process can require a residence time 720 shorter than a batch reactor because the microchannels' interfacial area is 32 times larger. The same reaction developed by batch reactor and microreactor showed a yield of 94.1% in 180 minutes against 95.8% in 1 minute, respectively (Santana et al., 2016). The process enlargement can be carried out by parallel or series arrangement of the micro/millireactor, by the modular scale-up concept, in combination with a slight increment in the channel dimensions (scale-out), as, for example, from micro to milliscale (Santana et al., 2018b; Zhang et al., 2017). These slight increment of the channel dimensions aims to increase the operating flow rate without missing the advantages of microscale's enhanced transport phenomena. This strategy can be performed for any type of microreactor, and each specific reaction process will have an optimal channel dimension. Microplants have been shown promising results, such as those obtained by Billo et al., (2015). He produced 2.47 L.min-1 of biodiesel by multiplying 14.000 microreactors. The plant had 35 manifolds with 8 modules, and each module with 50 microdevices. The whole fabrication process took 3 months (Billo et al., 2015). In addition to overcoming the macroscale plant issues aforementioned, the microscale configuration has advantages such as higher process security and control, large surface contact area, low shear stress, easy management of thermal exchange and reaction kinetics, easy maintenance, and repairability without completely stopping production (Budžaki et al.,

2017; Sun et al., 2008a, 2008b; Wirth 2013). However, strategies to intensify heat and mass transfer rates are the key factor to achieving high efficiency.

Contrary to the conventional processes, by using microdevices, the heat and mass transfer are more homogeneous without spots that can compromise the reaction development. One strategy is the use of micromixers and/or complex geometry to promote turbulent mixing spots or chaotic advection. Martínez Arias et al., (2012), evaluated the transesterification of castor oil and ethanol catalyzed by NaOH. The shape that fomented chaotic flows (Tesla-shape) favored turbulent mixing and resulted in higher conversion (92%). In comparison, the shape that fomented laminar flow (T-shape) resulted in lower conversion (73%). Hence, studying the fluid flow is essential to optimize the process, and it can be developed using Computational Fluid Dynamics (CFD) techniques. CFD provides complete and detailed process information for a given fluid flow. Variables such as velocity, density, viscosity, temperature, species concentrations, and reaction rates may be acquired for all computed points in the flow domain. Moreover, beyond CFD being used for process design, it can complement experiments in process analysis (Chung 2002). The use of microscale allied to CFD can also be adopted for biodiesel purification, considering that the separation of the reaction products is a fundamental key in the process feasibility. Thus, the microscale plant can efficiently contemplate the entire process of biodiesel production, and the purpose of this chapter is to provide a comprehensive summary of these technologies.

Literature Review

Biodiesel Synthesis in Microreactors

Biodiesel refers to a fatty acid alkyl ester (FAAE) obtained from renewable sources. The feedstock (i.e., canola oil, waste cooking oil, and soybean oil) reacts with short-chain alcohol (i.e., methanol, ethanol, and isopropanol) in the presence of a homogeneous or heterogeneous catalyst (i.e., sulphuric acid, potassium hydroxide, and lipases), resulting in FAAE. Two reactions routes are the most common: esterification (Reaction 1) by using free fatty acid (FFA), and transesterification (Reaction 2) by using triglycerides (TGL). High yields are achieved by optimizing feedstock properties, alcohol, catalyst, temperature, mixing, and reactor shape and size. The batch reactor is frequently used, but it takes several hours for high yield (Yeh et al., 2016). By using microreactors, it is possible to achieve superior efficiency in a short time. The same reaction developed by batch reactor and microreactor showed a yield of 94.1% in 180 minutes against 95.8% in 1 minute, respectively (Santana et al., 2016). According to Bannatham et al., (2021), the microscale process requires a residence time 720 shorter than a batch reactor because the microchannels' interfacial area was 32 times larger. The most significant gap of this process is the reactants' miscibility. It can be overcome by using high alcohol content and temperatures higher than 50ºC (Bonet 2014; Silva et al., 2015) or increasing the interfacial surface (i.e., ultrasound, supercritical conditions, co-solvents, and microscale reactors). Microreactors have a high surface area relative to the volume (Kobayashi et al., 2009), easier control of operating conditions, and better heating performance, i.e., superior homogeneous distribution (no heating spots) than macroscale devices. Moreover, microreactors processes can be quail and quantitatively optimized mainly by microdevices design, operating conditions (i.e., temperature, molar ratio, and flow rate), and catalyst.

$$R^1\text{-CO-OH} + R^2CH_3 \xrightleftharpoons{\text{Catalyst}} R^1\text{-CO-O-}R^2 + H_2O$$

FFA   Alcohol   FAAE   Water   (1)

$$\begin{array}{c} R^1\text{-CO-O-}CH_2 \\ R^2\text{-CO-O-}CH \\ R^3\text{-CO-O-}CH_2 \end{array} + 3\,R^4CH_3 \xrightleftharpoons{\text{Catalyst}} \begin{array}{c} H_2C\text{-OH} \\ HC\text{-OH} \\ H_2C\text{-OH} \end{array} + \begin{array}{c} R^1\text{-CO-O-}R^4 \\ R^2\text{-CO-O-}R^4 \\ R^3\text{-CO-O-}R^4 \end{array}$$

TGL   Alcohol   Glycerin   Mixture of FAAE   (2)

Microdevices Design

Different designs for microreactors have been evaluated to carry out biodiesel reactions. The major aspect of a higher yield is to obtain a better reactants' contact and mass transfer. Lukić and Vrsaljko, (2021), observed that the two reactants form segmented interspersed flow and the reaction occurs between both reactants' borders. Thus, smaller reactants' segments result in a larger conversion. Two artifices are frequently used for that: (1) reduction of the channel cross-sectional area and (2) use of complex channels geometry.

Firstly, smaller cross-sections increase the reactants' area contact, although it also accentuates the pressure drop. In contrast, a larger cross-sectional area, which requires low pressure, results in a lower conversion. As observed by Guan et al., (2010), the reaction yield obtained by microchannels with 0.96 mm and 0.46 mm is similar (89.2% and 91.7%, respectively) for a residence time of 252 s. In contrast, for a residence time of 112 s, a significant difference in the conversion was observed (43% and 80%, respectively).

Secondly, the microdevice geometry influences the reaction yield because it improves the reactants mixture. The mixing improvement often occurs by a micromixer adapted to the microreactor inlets or by turbulent mixing spots, or chaotic advection, inside the reactor developed by complex microchannels designs. The micromixer design was evaluated by Rahimi et al., (2016), for soybean and methanol transesterification catalyzed by NaOH. Three designs with confluence angles of 45°, 90°, and 135°, respectively, were compared and the first one resulted in higher conversion (98.8%). Martínez Arias et al., (2012), evaluated the transesterification of castor oil and ethanol (1:9 molar ratio) catalyzed by NaOH (1 wt%) at 50°C. The shape that fomented chaotic flows (Tesla-shape) favored turbulent mixing and resulted in higher conversion (92%). In comparison, the shape that fomented laminar flow (T-shape) resulted in lower conversion (73%). Moreover, designs that fomented turbulent mixing require low alcohol content (Martínez Arias et al., 2012). Turbulent mixing yielded a similar conversion for 1:9 and 1:24 molar ratios (92% and 97%, respectively). At the same time, a higher difference was observed for laminar flow (73% and 94%, respectively). Rahimi et al., (2014), searched for the use of different microdevices designs for soybean oil and methanol (1:9 molar ratio) transesterification catalyzed by KOH (1.2% w/wt) at constant temperature (~60°C). A circular microtube with a T-shaped micromixer achieved a conversion of 89% with a residence time of 180 s (Rahimi et al., 2014). In comparison, an optimized zigzag micro-channel reactor with a T-shape inlet with a three-way junction achieved a conversion of 99.5% with a residence time of 28s (Wen et al., 2009).

Moreover, although unexplored for microscale, the use of microwave, ultrasound and coil, for example, could improve the mass transfer (Bucciol et al., 2020; Chipurici et al., 2019; Thangarasu et al., 2020). Palm fatty acid distillate (PFAD) feedstock consisting of >90% free fatty acid and methanol (1:9 molar ratio) catalyzed by sulfonated glucose (2.5 wt%) was carried out in an oscillation flow microreactor (OFR) (60°C, residence time of 50 minutes, and oscillation frequency of 6Hz) resulting in a biodiesel yield of 94% (Kefas et al., 2019). The OFR is a continuous oscillatory flow tubular reactor adaptable for homogeneous and heterogeneous catalysts (Eze et al., 2013; Kefas et al., 2019). A mixture of oils and methanol (1:9 molar ratio) catalyzed by KOH was carried out in a helical coil microreactor by single flow (SFHR) and reverse flow (RFHR). At the same operational conditions (2g of catalyst, 65°C, and residence time of 5 minutes), both reactors resulted respectively in 91% and 99% of biodiesel production. The difference occurred because the RFHR operates by a reverse turn coil. It results in a secondary flow with a dipole-like velocity field development; further, the flow near the inner surface of the coil generates large eddies. Both behaviors help to improve the reactants mixture, resulting in an 8-9 times higher reaction efficiency than a batch reactor (Gupta et al., 2019). Therefore, the microdevices design can be explored in different ways to achieve high efficiency by focusing, for example, on mild operational conditions and low-cost catalysts.

Operational Conditions

Esterification and transesterification usually require high temperature and alcohol content to promote efficient mixing. The conventional processes, by using acid and alkali, almost use temperatures higher than 100°C (Georgogianni et al., 2009; Stacy et al., 2014; Tabatabaei et al., 2019). On the contrary, using a microreactor, temperatures higher than 50°C are frequently desirable, but lower alcohol (3 mols of alcohol for 1 FAAE mol) content is required (Billo et al., 2015; Guan et al., 2010; Martínez Arias et al., 2012; Mohadesi et al., 2020a; Rahimi et al., 2014; Wen et al., 2009). However, the microdevices design impacts both operating conditions. Martínez Arias et al., (2012), evaluated the temperature influence in different microdevices shapes for castor oil and ethanol transesterification. The Tesla shape, which fosters turbulent spots and better mixing, achieved 75%, 93%, and 93% for 30, 50, and 70°C, respectively. In contrast, T-shape, which fosters laminar flow and poor mixing, achieved 66%, 73%, and 88% for 30, 50, and 70°C, respectively. These results are in agreement with Lukić and Vrsaljko, (2021), study. They observed the formation of segmented interspersed flows between both reactants. The reaction occurred between both reactants' borders, and the higher the temperature, the smaller each reactant segment. At a temperature higher than the alcohol boiling point, the mixture stops being a liquid-liquid interface to becomes a gas-liquid interface, resulting in convective mixing and consequently improving the FAAE conversion. Therefore, microdevices that foment better mixing require temperatures higher than 50°C with liquid-liquid interfaces. Meanwhile, microdevices that cannot develop a good mix require temperatures higher than the alcohol boiling point to achieve higher conversions.

The molar ratio required for esterification is 1 mol of FFA for 1 mol of alcohol (Reaction 1). For transesterification, 1 mol of TGL is required for 3 mols of alcohol (Reaction 2). However, the alcohol excess is usually considered, as it helps to obtain a superior mixing between the reactants. By using acid or alkali, the traditional processes require 50-300 mols of alcohol for each FFA/TGL mol (Gopi et al., 2022). In contrast, microdevices, which foment better mixture, require a range between 3 to 30 molar alcohol for 1 TGL or FFA (Billo et al., 2015; Guan et al., 2009a; Martínez Arias et al., 2012; Mohadesi et al., 2020a; Rahimi et al., 2014; Wen et al., 2009). For example, castor oil transesterification resulted in a conversion of 93% for 1:9 molar ratio and a constant conversion value of 97% for 1:12, 1:18, and 1:24 molar ratio by carrying out the reaction in a reactor shape that improves the reactant mixing. In comparison, a result of 76%, 83%, 92%, and 93% were achieved for 1:9, 1:12, 1:18, and 1:24 molar ratio, respectively, by carrying out the reaction in a reactor shape that promotes lower reactant mixing (Martínez Arias et al., 2012). Even so, the microscale process requires a lower alcohol content than the conventional processes.

Catalyst

Homogeneous catalysts, such as strong acid or alkalis, are mostly used for biodiesel production on microscales and industrial scales (Gopi et al., 2022). Strong acids can be used for esterification and transesterification. However, the reaction is slow, and the reactors demand a non-corrosive material finish. Anyhow, damages to the reactor can occur and need to be repaired. For an industrial scale reactor, it is necessary to stop a large reaction volume. At the same time, on a microscale, the process continues, requiring the replacement of a few units by cheaper maintenance. The alkalis catalyze just transesterification; however, by a faster reaction than a strong acid. Moreover, the alkali reaction requires free water feedstock; otherwise, saponification may occur as described by reaction (3). Considering that saponification may occur, a large volume of water and reactants in an industrial-scale reactor is necessary to solve it. Although, in a microscale reactor, it would probably happen in sparse units, and the problem would be solved with less volume of reactants and water, without stopping production.

$$R^1-O-\overset{\overset{O}{\|}}{C}-R^2 + KOH \xrightarrow{H_2O} R^1-OH + K^+\,{}^-O-\overset{\overset{O}{\|}}{C}-R^2$$
Ester　　　Alkali　　　　Alcohol　　　Soap　　　　　　　　(3)

Heterogeneous catalysts have advantages such as easier catalyst recovery and reuse, and higher purified product (Ebiura et al., 2005; Essamlali et al., 2017; Xie et al., 2006; Zabeti et al., 2009; Zhang et al., 2010). Gojun et al., (2021), achieved 94% of biodiesel conversion from sunflower oil and methanol (1:10 molar ratio) catalyzed by a lipase from *Thermomyces lanuginosus* (723.34 ± 19.19 U/mg) in a microdevice (40°C and residence time of 20 minutes). The oil+enzyme and methanol entered the microreactor by a T-shaped micromixer. Mohadesi et al., (2020b), achieved 97% of biodiesel conversion from WCO and methanol (1:2.25 molar ratio) catalyzed by KOH/Clinoptilolite (8.1 wt%) in a micro-tube reactor (65°C and residence time of 13.4 minutes). The oil+catalyst and alcohol entered the reactor by a T-shaped micromixer. Mohadesi et al., (2020a), achieved 97% of biodiesel conversion from WCO and methanol (2.15:5 molar ratio) catalyzed by kettle limescale deposit (15 wt%) and acetone as co-solvent (13.95 wt% methanol) in a microreactor (60°C and residence time of 12.5 minutes) with T-shaped micromixer entered.

Low-cost material as a catalyst has been evaluated. Mohadesi et al., (2021), achieve 99% of biodiesel conversion from WCO and methanol (1:2.5 volume ratio) catalyzed by cow bone (8.5 wt%) in microdevice (63°C and residence time of 60 s). The WCO+cow bone and the methanol entered to microreactor by a T-shaped micromixer into a microchannel's diameter of 0.8 mm. Pavlović et al., (2021), achieved 51% of biodiesel conversion from sunflower oil and methanol (1:2.5 volume ratio) catalyzed by chicken bone (10 wt%) in microdevice (60°C and residence time of 10 minutes). The oil+chicken bone and methanol entered to microreactor by a T-shaped three-way micromixer into a microchannel's diameter of 0.8 mm.

Packed micro-channel reactors for continuous biodiesel production have also been explored. Palm oil and methanol in the presence of calcium oxide pretreated with methanol as catalyst resulted in a conversion of 99% by using 65°C, 1:24, and residence time of 8.9 min. The catalyst was stable for an operating time of more than 24 hours (Chueluecha et al., 2017). According to Mohadesi et al., (2020a), residence time, temperature, and molar ratio are the most important operational conditions for biodiesel production by heterogeneous catalysis in microdevices.

Enzymes, frequently lipases, are one of the most common heterogeneous catalysts which have been evaluated for biodiesel production. They have been used in microreactors by edible and non-edible oils (Budžaki et al., 2017). This process has the potential for sustainable industrial biodiesel production (Budžaki et al., 2017). However, it presents three challenges: efficient immobilization, catalyst cost, and reactants mixing (Budžaki et al., 2017; Rossetti 2018). By using a microscale reactor, the mild operating conditions reduce the enzyme demobilization and loss. However, low flow rate or poor mixing can affect the enzyme efficiency, considering that glycerol can impregnate the channels and catalyst surfaces, therefore requiring additional removal operation. For an industrial scale reactor, a process involving a large volume of reactants is necessary to solve it. Although, in a microscale reactor, it would probably occur in sparse units, and the problem would be solved with less volume of reactants without stopping production.

Biodiesel can also be produced by non-catalyst methodologies, such as supercritical conditions (Kusdiana & Saka 2001; Qadeer et al., 2021; Saka & Kusdiana 2001), electrolysis (Fereidooni et al., 2021; Guan & Kusakabe 2009), or plasma (Buchori et al., 2016; Oliveira Palm et al., 2022). An advantage of the non-catalyzed reaction is that is not required a catalyst recovery, reducing cost and wastewater production. However, these technologies are unexplored by microdevices. Although it can be successfully applied as observed by Akkarawatkhoosith et al., (2021), who achieved 97% of rice bran oil conversion using supercritical conditions (360°C, 1:11 oil to alcohol molar ratio and residence time of 35 minutes) in a microreactor.

Biodiesel Purification in Microdevices

Transesterification of oils with short-chain alcohol leads to the formation of biodiesel and the co-product glycerol, in addition to the excess alcohol. The separation of these three compounds is a fundamental key in the process's feasibility. In addition, the reactor downstream can also contain impurities from feedstock, unsaponifiable materials, residues from catalyst and water (Gopi et al., 2022). Post-treatment stages of biodiesel are available, including distillation, membrane filtration, liquid-liquid extraction, and wet and dry washing, being the last two the most common methods currently used to achieve the biodiesel standards (Fonseca et al., 2019).

Usually, for traditional batch processes, the reacted mixture follows a gravitational separation process due to the density difference of the resulting phases. This separation also experiences an emulsion formation, making the separation relatively slow and hard to accomplish (Tiwari et al., 2018). In microscale, the driving force of density difference is very reduced, and then gravitational separation is not possible. In contrast, other driving forces can be intensified in microscales, such as the centrifugal forces and secondary vortex generation (Dean flow) employed in inertial focused devices to separate biological particles (Nasiri et al., 2020).

Process intensification strategies include the use of integrated operations, e.g., reaction and separation in the same equipment. Membrane reactor appears as an interesting alternative for biodiesel production (Gao et al., 2017; Reyes et al., 2012; Xu et al., 2015; Maia et al., 2016). The two-phase membrane reactor performs the transesterification and separation of the biodiesel in a single continuous operation (Gao et al., 2017). The continuous separation of the products from the feed stream (reactants mixture) allows a high mass transfer rate between the immiscible phases (Baroutian et al., 2011; Cao et al., 2009). It is a flexible alternative since the most usual transesterification routes (i.e., acid, alkali, and enzymatic, homogeneous, or heterogeneous media), can be coupled to the microfilter membrane (Atadashi et al., 2012; Baroutian et al., 2012). These characteristics produce an oil-free biodiesel stream at the membrane reactor outlet, thus resulting in simple downstream processing and lower energy demand (Shuit et al., 2012; Kiss et al., 2012).

There are few studies in biodiesel separation and purification in microdevices, such as the use of Liquid-Liquid extraction performed by Crawford et al., (2008). The authors used an integrated reactor-extractor system made by a Syrris 250 µL microchannel reactor arranged in series with a Flow Liquid-Liquid Extraction (FLLEX) module, which was composed by a polytetrafluoroethylene (PTFE) membrane (Kralj et al., 2007). The continuous separation of glycerol from biodiesel was achieved by the introduction of water in the stream, where the hydrophobic biodiesel phase was able to flow across the membrane, in contrast to the hydrophilic aqueous/glycerol phase. An in-line mass spectrum system was coupled to the micro-plant, allowing the biodiesel quantification in real-time. A residence time of 2.5 minutes was enough to provide a complete conversion of the vegetable oil at ambient temperature.

Recently, Bacic et al., (2021), proposed an integrated continuous process for biodiesel synthesis and purification by using Deep eutectic solvents (DESs) with lipase catalyst. The performance was assessed for batch system and microreactor. The microsystem was composed of a Y-shaped input with a tubular channel of an inner diameter of 500 μm and 1.2 m long made by polytetrafluoroethylene (PTFE). At the optimal conditions, the batch reactor provided 43.5% of biodiesel yield and extraction of 99.54% of glycerol after 2 h (w_G = 0.027% wt of glycerol); however, it did not achieve the international standards for glycerol content. The microsystem provided a similar performance, 45.33% of biodiesel yield and 99.56% of glycerol removal, for a residence time of 2 h, w_G = 0.019% wt of glycerol content, achieving the international standards (w_G = < 0.02%). The authors concluded that the microsystem can be considered a starting point for further development of integrated production/purification biodiesel synthesis in a single stage.

The excess alcohol also can be recovered in the downstream stages. Santana et al., (2017a), designed and fabricated a micro-heat exchanger (MHE) to perform the excess ethanol evaporation from the biodiesel stream. The MHE consisted in a multichannel (15 channels of 500 μm of hydraulic diameter) microdevice with a Peltier plate to provide the heat required for ethanol evaporation.

The microdevice was manufactured by poly(dimethylsiloxane) (PDMS) using soft lithography. The excess ethanol was evaporated and recovered at one of the two outlets of the device. The operating variables' effect in the ethanol recovery were assessed by ranging the ethanol/biodiesel molar ration from 2 to 11, temperature from 80°C to 120°C, and flow rate from 0.1 mL/min to 1.2 mL/min. The superior performance of the MHE was a recovery efficiency of 82% for ethanol/biodiesel molar ratio of 2, at 100°C and 0.6 mL/min for a single pass through the microdevice. The MHE design was also evaluated by numerical simulations using CFD by Silva Jr. et al., (2017). These results testify to the potential usage of the MHE in continuous modular plants.

Introduction to Computational Fluid Dynamics (CFD) Concepts

Computational Fluid Dynamics

Computational Fluid Dynamics (CFD) is the area of fluid mechanics based on applying computational tools to solve mathematical models that describe fluid flows. It is a heavily used approach to analyze two- or three-dimensional flows, whose analytical solution is too difficult to obtain. Hence, computers are used to speed up this process through numerical solutions.

There are three basic laws that determine a fluid flow: mass conservation, Newton's second law, and energy conservation. Therefore, flows are represented in terms of mathematical equations described in partial differential equations. CFD techniques are applied in order to transform these equations, which represent a continuum behavior, in algebraic equations, interpretable and solvable by computers by appropriate algorithms (Bhatti et al., 2020).

Advantages of Applying CFD in Process Analysis

Computational fluid dynamics techniques are nowadays well established and applied in many technology fields. Some heavy CFD users are manufacturers of aircraft, cars, ships, and turbomachinery. Besides these, such a design approach has been adopted to solve problems in oceanography, meteorology, astrophysics, and others (Moukalled et al., 2016).

Cost-saving is by far the main advantage of applying CFD in process and equipment design when compared to experimental investigations. Studying a process through a proper set of simulation runs based on a validated model saves financial, human, and material resources than the same set of experiments in a pilot plant. Furthermore, undetermined and human errors that may influence lab tests are eliminated, while well-conducted numerical solutions always tend to provide accurate results.

CFD analysis also provides information difficult to obtain via experiments. On the one hand, it is possible to obtain flow details as velocity field throw particle image velocimetry (PIV) with resolution limitations. On the other hand, a simulation provides a velocity field which resolution only depends on the available computational resources. Besides that, real-time experimental data on reactive flows is not easily accessible. Although techniques like Raman spectrometry allow such data acquisition, they are limited by few samples for a given flow domain. On the contrary, CFD analysis provides results as species concentration fields, i.e., throughout the flow domain, in real-time when considering a transient process (Anderson, 1995).

Summarizing, CFD provides complete and detailed process information for a given fluid flow. Variables such as velocity, density, viscosity, temperature, species concentrations, and reaction rates may be acquired for all computed points in the flow domain. Furthermore, not only CFD can be used in process design but it also can complement experiments in process analysis (Chung, 2002).

CFD Analysis Workflow

The steps to analyze a problem using CFD are represented in the flowchart shown in Figure 1.

The starter step when developing a CFD model is the observation of a given process. This consists not only of phenomena investigation but also of assumptions survey, based on observer's experience. The next step is physical modeling, which consists in establishing a control volume for the fluid flow by imposing control surfaces. This basically means representing the flow domain using a geometry, usually developed in a CAD software.

Thereon, a mathematical modeling is applied in order to represent how the flow parameters vary throughout the domain. This consists of applying basic conservations laws to the control volume, so there are dependent field variables that describe the flow. This step results in a set of partial differential equations (PDEs) where, e.g., pressure and velocity are dependent field variables, and independent field variables are dimensions and time in a transient problem.

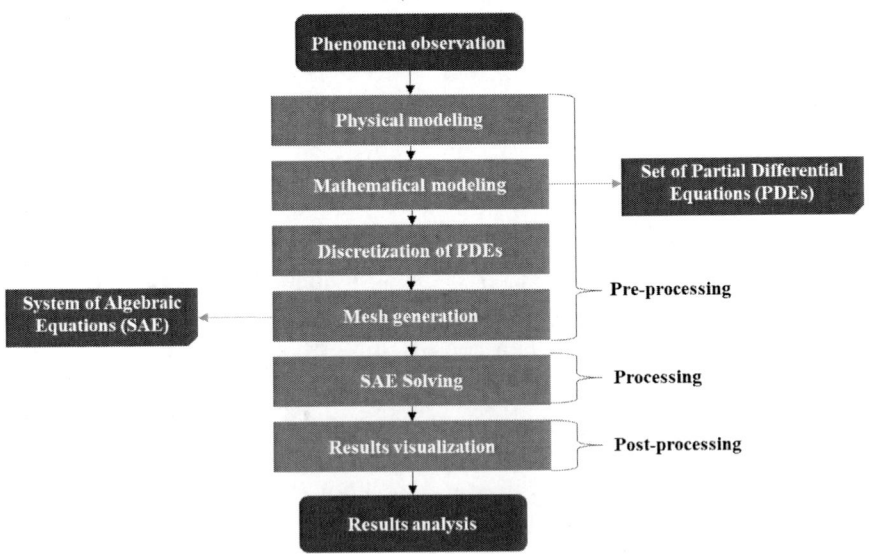

Figure 1. CFD analysis basic steps.

Subsequently, PDEs are discretized. This step is needed for adequate implemented algorithms to solve the fluid flow through a digital computer. Such discretization step can be done using methods such as finite differences

(FDM), finite elements (FEM), or finite volumes (FVM). Conversely, mesh generation means the spatial discretization of the physical domain. In this step, an inherently continuous geometry is transformed into a domain composed of a finite number of points, elements, or cells, the so-called mesh or grid. PDEs discretization and mesh generation then result in a system of algebraic equations (SAE) for each domain element, where the field variables are now discrete variables. The set of PDEs that was then applicable to the whole domain is now represented by a SAE for each mesh element (Moukalled et al., 2016).

The steps described above are called pre-processing steps, i.e., the needed steps to prepare a fluid flow case for a computer algorithm to solve it by transforming a continuous into a discrete domain. The simulation run is the processing step. It consists of the SAE solution for each element of the domain. There are many well-established numerical methods which allows the solution of linear and non-linear problems. Such nonlinearities usually appear in reacting flows, e.g., when reaction rates are described by second or high order kinetics. Once the numerical simulation converges, the subsequent steps are post-processing. This can be done by an adequate software to obtain flow details graphically for any computed field variable (Moukalled et al., 2016). A few more details about pre-processing, processing, and post-processing steps of a CFD analysis are given in the subsequent sections.

Physical Modeling

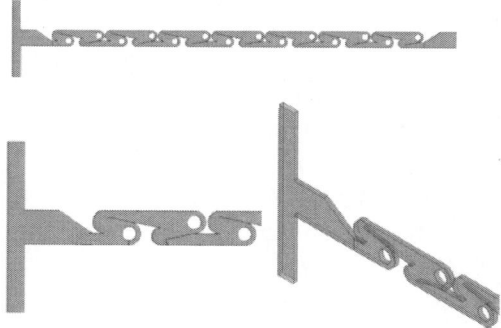

Figure 2. Design of the micromixer "Elis," developed for enhanced fluid mixing and organic reactions.

In CFD analysis, physical modeling is the development of a geometry that represents the control volume to be analyzed. There are commercial and free CAD software, such as SALOME, Gmsh, SketchUp, and others that allow geometry design. This step is critical both in process design and analysis. In a process analysis, it is important to assure that the control volume represents, e.g., the needed length for a fully developed outflow. In a process design, geometry design is determinant to assure, e.g., reactants mixing effectiveness in a microreactor. Figure 2 represents a micromixer which geometry was developed in such a way that reactants mixing is enhanced by obstacles (Santana et al., 2021).

Mathematical Modeling

There are two different approaches to describe a fluid flow. The Lagrangian approach is based on describing the path of a given particle throughout a flow domain. On the other hand, the Eulerian approach is based on describing how field variables vary throughout a given domain where all points are specified. From this point on, the fluid dynamics concepts are described based on the Eulerian approach.

Developing a mathematical model consists of applying the conservation equations for the control volume that represents a given flow domain. The starter step is to apply the proper conservation equations to an infinitesimal control volume. Thus, such equations are represented in a set of PDEs. The solution of these PDEs gives fluid flow details by fields of velocity, pressure, density, etc.

In a three-dimensional flow, independent variables are scalar variables, such as dimensions and time. The field variables to be calculated can be pressure and velocity. In this, as in all other cases, there are vector field variables, as velocity, and scalar field variables, as pressure, by definition (Moukalled et al., 2016).

The simpler three-dimensional flow to be solved can be described by a set of two differential equations, those for mass and momentum conservation. The mass conservation equation is a scalar equation given as:

$$\frac{\partial \rho}{\partial t} + \nabla \cdot (\rho \vec{u}) = 0 \qquad (4)$$

where ρ is fluid density and $\vec{u}$ is velocity. Equation 4 is also called the continuity equation, since its formulation is based on the *continuum* hypothesis. This assumption considers that all flow properties vary continuously in space and time. The first term on the left side of the equation represents the mass variation in the control volume, while the second term represents the balance of mass flux across control surfaces.

The momentum conservation equation is derived from Newton's second law of motion. When this law is applied to a case of incompressible, isothermal, and laminar flow of a Newtonian fluid, the resulting equation is known as the Navier-Stokes Equation. This is a vectorial equation given as:

$$\frac{D\vec{u}}{Dt} = \rho\vec{g} - \nabla P + \nabla \cdot [\mu(\nabla\vec{u} + {}^t\nabla\vec{u})] \qquad (5)$$

where the term on the left side of equation represents the material derivative, which can be understood as the advective term. On the other hand, $\nabla \cdot [\mu(\nabla\vec{u} + {}^t\nabla\vec{u})]$ is the momentum diffusion term for a Newtonian fluid, ∇P is the pressure gradient, which gives the flow direction, and $\rho\vec{g}$ is the term of gravitational force, the most usually field force present in flow cases.

Here, we presented the concept of advection and diffusion of transport variables. Advection is the transport due to a bulk fluid motion, and diffusion is the transport mechanism due to molecular interaction. Additional terms would be needed to describe momentum transport if there was turbulent shear stress (Bird et al., 2006).

In chemical processes, it is also critical to describe species conservation. Biofuel synthesis eminently involves the transfer, generation, and consumption of chemical species. Thus, we must emphasize that, as momentum transport, species transport is represented by equations composed by advective and diffusive terms. Generally, conservation of a given species can be described as:

$$\frac{\partial C}{\partial t} + \nabla \cdot (\vec{u}C) = \nabla \cdot [\mu(\nabla C)] + r \qquad (6)$$

where is $\frac{\partial C}{\partial t}$ the concentration of a species variation with time, $\nabla \cdot (\vec{u}C)$ is the advective term, $\nabla \cdot [\mu(\nabla C)]$ is the diffusive term, and r represents the balance of this species concentration when there is a reactive flow, expressed by the reaction kinetics, given as:

$$r_i = r_{generation} - r_{consumption} \qquad (7)$$

These governing equations are naturally coupled, so all equations need to be solved simultaneously. In transient processes, the obtained solution express variables that vary not only with space, but also with time.

The relation between physical and mathematical models is established by boundary conditions. They are constraints presented in the problem adopted at the control surfaces, always represented as equations. Such equations are requirements for solving the set of PDEs (Chung, 2002).

Discretization of a Mathematical Model

A digital computer can solve a mathematical model problem only if the set of PDEs are described into systems of algebraic equations. The three main methods used to discretize equations in a CFD model are: finite-difference (FDM), finite-element (FEM), and finite-volume methods (FVM).

Basically, the finite-difference method consists of assuming a point in space where the *continuum* hypothesis can be considered for the original set of PDEs, and then these equations are turned into discrete equations, the so-called finite-difference equations. It is a quite suitable discretization option if there is a regular domain shape, like rectangular or box-shaped. On the other hand, this method does not handle well with variables discontinuities due to geometry. Also, it is too difficult for implementation in complex shapes.

Finite-element methods are based on subdividing the physical model in small parts of simple shape, that form a mesh of finite elements. The set of PDEs are formulated not only for the entire domain, but for each element. Thus, this method consists of approximate field variables in simple low-degree polynomial functions, such as linear or quadratic. This procedure results in a local approximation into an SAE for each element. When this formulation is extrapolated for all domain elements, then an SAE is obtained, represented by a sparse matrix that can be solved through any well-known sparse matrix solver. This method allows to increase results accuracy in specific regions of a domain, as corners, by increasing the number of elements, i.e., refining the mesh. However, finite-element methods require relatively advanced mathematical expertise for their implementation, and transient cases are even more complex (Pulliam et al., 1999).

Finite-volume methods are based in applying conservation laws to elements of a domain called cells. Thus, finite-volume methods are similar to finite-elements since both are based on dividing a geometry in small shape elements, although the adopted approaches over governing equations are quite different. In FVM, a flux variable that enters in a cell's face has to leave the other side's face of the same cell. Therefore, implementing these methods results in a set of flux conservation equations defined for mean variables at the cells. Since most of governing equations of a fluid flow are based on conservation laws, FVM is successful to solve such problems. As in FEM, this method provides more accurate results if a mesh is locally refined (Hirt and Nichols, 1981; Pulliam, et al., 1999).

Mesh Generation

Meshing consists of discretizing the physical model in space, i.e., generating a discrete representation of a fluid domain geometry. This mesh unfolds which are the points to which the equations will be solved, and which points belong to the interior and to the boundaries of the domain. A mesh strongly influences convergence rate or even the lack of convergence, accuracy of the results, and time required for a computer to solve a simulation. Hence, CFD designers attempt mesh quality parameters towards reasonable simulation results.

Volume elements of a mesh are usually called cells, which constitute each control volume of a 3D grid. These elements may be tetrahedrons, hexahedrons, or polyhedrons. Vertices are called nodes, and a centroid is a point referring to the center of a cell, to which flow properties are computed during a simulation run. Lines are called edges, which constitute the boundaries of a face. So, a face is a boundary of a cell. Nodes, faces, and cells may be part of a zone. Finally, a flow domain is made of nodes, faces, and cell zones (Bakker, 2006). The described mesh terminology is represented in Figure 3.

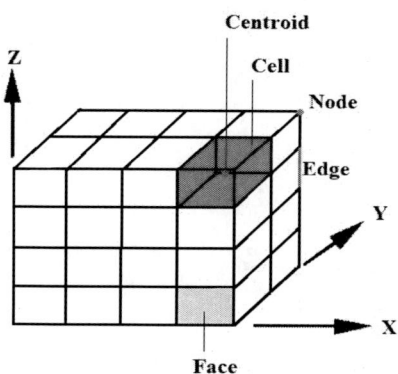

Figure 3. Representation of mesh terminology concepts.

Regarding to the mesh types, there are structured and unstructured meshes. A structured mesh is one where there is cell indexing in order to locate nearing cells of one for which the problem is being solved. Thus, structured meshes can only be applied in simple geometries, so that all cells of a 3D domain may be addressed by three index variables, e.g., i, j, and k. An unstructured mesh is composed of elements arranged without indexing, so there is no limitation concerning geometry shape (Moukalled et al., 2016). However, this type of mesh implies in memory and CPU overhead due cells referencing. There are a few mesh quality parameters, as mesh density, skewness, smoothness, and aspect ratio, whose definitions may be well comprehended in the literature.

Simulation Runs

As above mentioned, a simulation run consists of solving SAE for each element of discretized flow domain. There are well-known established numerical methods for this purpose implemented in commercial and free software. A few examples of available CFD commercial software are COMSOL, Ansys CFX, Ansys Fluent, and Siemens StarCCM+. Open-source CFD solutions are also available as OpenFOAM.

Post-Processing

Results of a simulation run can be visualized using a proper post-processing software. Field variables can be expressed graphically so that process variables can be seen for the entire flow domain. Figure 4 expresses how the oil mass fraction varies in a biodiesel synthesis in a simulation run applied to microreactor design from Figure 2 (Santana et al., 2021).

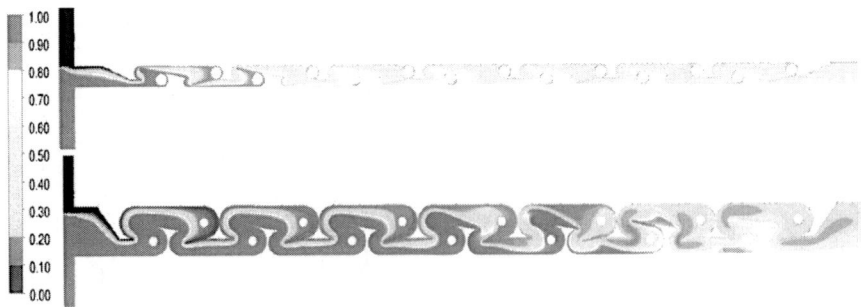

Figure 4. Vegetable oil mass fraction distributions for different operating conditions simulated in Elis microreactor for biodiesel synthesis.

Numerical Simulation of Biodiesel Synthesis in Microreactors

Biodiesel production has been increasing with the demand of reducing fossil fuels consumption. This way, engineers also have been investigating how to design a rapid and safe production process to supply such demand. Since biodiesel synthesis is a multiphase transesterification reaction, slow mass transfer impacts overall reaction time and continuous reactors have been developed to overcome this issue (Quiroz-Pérez et al., 2019). To implement this production process at industrial scale, mass and heat transfer must be enhanced. In this context, microfluidic devices provide some major advantages like a higher area to volume ratio, which enhances mass and heat transfer and smaller reaction times (Zhang et al., 2006).

Since CFD modeling is a powerful tool in process design, the improvement of biodiesel production via microreactors has been explored through simulation in recent years. One of the first CFD investigations was conducted by (Han et al., 2011), which studied biodiesel synthesis in a capillary microreactor by the transesterification of soybean oil and methanol. Research on biodiesel production by experiments and numerical simulations were also conducted using a tubular microreactor (López-Guajardo et al., 2017). However, through intense research, it was observed that one of the key parameters that influence oil conversion is the reactants mixing. Usually, increasing flow velocities provide an efficient mixing degree in microreactors, but this can also be obtained by modifying the reactor design.

Further investigations indicated the effects of static obstacles in oil conversion. It was found that a reactor design with a cross-shape promoted the highest mixing index, since internal obstructions in the microchannel split and recombined the flow stream, which enhances mixing between chemical species (Santana et al., 2017b).

Through CFD results such as velocity profile, as seen in Figure 5, it was found that a design with obstacles in an alternating pattern result in a level of flow perturbation that increases the contact surface area between reactants. It was also found that baffles promote the flow direction changes and the formation of vortexes, resulting in higher mixing, thus enhancing mass transfer and oil conversion, as in Figure 6.

Mixing enhancement in a microreactor for biodiesel production has also been studied by (Mohd Laziz et al., 2020). This investigation, based on experiments and simulation, resulted in a continuous production of biodiesel from palm oil and methanol with a reaction time of 40 s and oil conversion of 98.6% using a slug flow of methanol.

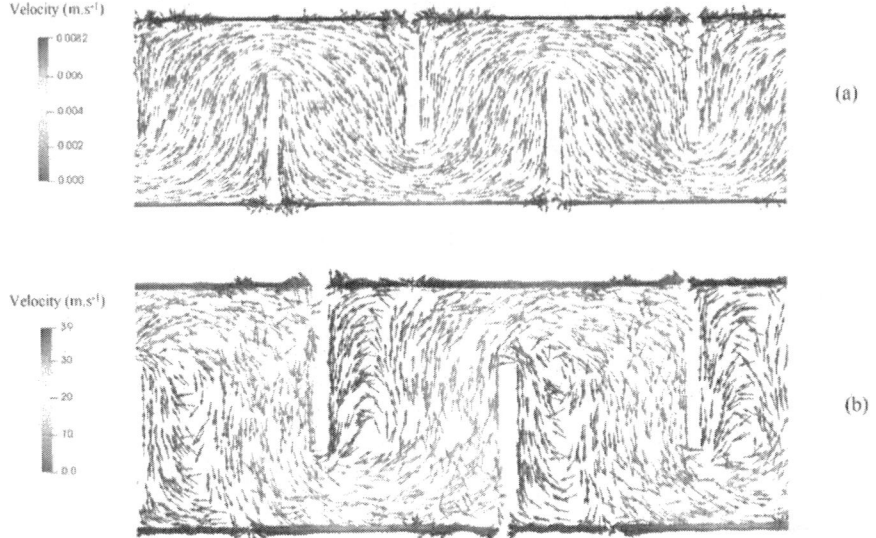

Figure 5. Velocity profiles in a microreactor with deflecting obstacles, or baffles, in different configurations of biodiesel synthesis: (a) Re = 0.01; (b) Re = 100. Reprinted from International Journal of Multiphase Flow, 132, Marcos R.P. de Sousa, Harrson S. Santana, Osvaldir P. Taranto, Modeling and simulation using OpenFOAM of biodiesel synthesis in structured microreactor, 103435, (2020), with permission from Elsevier.

Research has also been focused on approaches alternative to alkaline catalysis in biodiesel synthesis in microreactors. Some studies are based on ultrasound technology, supercritical solvents or the use of catalytic enzymes for the transesterification process (Akkarawatkhoosith et al., 2020; Basiri et al., 2016). These approaches result in lower energy consumption, while avoiding the formation of undesirable products. (Gojun et al., 2019) validated a model in COMSOL to investigate biodiesel reaction yield enhancement and found that fatty acids methyl esters yield was 30% higher in a microreactor than a batch reactor.

Fundamentals of Scale-Up and Scale-Up of Microfluidic Devices

Microfluidics systems present high yield and selectivity due to the reduction in mass and heat transfer resistances inherent from the higher surface to volume ratio, allowing superior process performance in short times regarding traditional macroscale equipment (Wirth 2013; Whitesides 2006). Furthermore, microdevices consume lower amounts of reactants and samples and can also be manufactured by 3D printing, reducing the costs and facilitating the design optimization and validation procedures.

Despite these advantages, the main issue lies on the very reduced throughput of the microdevice regarding the industrial scale demand. According to Zhang et al., (2017), to make feasible the use of microdevices, some challenges must be overcome, such as the use of thousands of microreactor units, the guarantee of uniform flow distribution among these units, and the total costs of fabrication of the microreactors and the assembly of the microchemical plant. Accordingly, the scale-up strategy is a fundamental point to make feasible the introduction of microfluidic devices in industrial processes.

Dr. Björn Mathes, in his article "Future Production Concepts in the Chemical Industry" argues that the scale-up of micro and millidevices can be employed to overcome the disadvantages of batch processes. In order to establish a batch process, there are some tests in pilot plants considering the adequacy of chemical reaction performance on a large scale. This time of process development and adaptation is even more critical in the pharmaceutical industry, which usually needs fast responses to the market demand, for example, in a pandemic. In addition, the continuous production using scaled-up micro and millidevices allows a relative reduction in reactants amount due to the higher yield, easier handling and automation, and low operational costs.

The scale-up can be carried out by the parallel or series arrangement of the micro/millireactor, by the modular scale-up concept, in combination with a slight increment in the channel dimensions (scale-out), such as, for example, from micro to milliscale (Zhang et al., 2017; Santana et al., 2018). These slight increment of the channel dimensions aims to increase the operating flow rate without missing the enhanced transport phenomena advantages from microscale. This strategy can be performed for any type of microreactor, and each specific reaction process will have an optimal channel dimension.

Vankayala et al., (2007), developed Falling Film Microreactors (FFMR) for the oxidation of organic compounds and CO_2 absorption with sodium hydroxide solution. The study evaluated the channel dimensions effects in the reaction performance. One microreactor has 16 channels of 1200 μm x 400 μm (width x depth), while the other configuration was a 32 channels reactor with dimensions of 600 μm x 300 μm (width x depth). The superior yield was noticed for the larger microreactor, a surprising result, since both reactors possessed the same wetted area.

Lin et al., (2021), applied the scale-out process in a liquid-liquid microextractor and obtained an extraction efficiency above 90% in 10 min. The applied strategy was effective in overcoming the low production rate. Kang and Tseng, (2007), developed a micro-heat exchanger and evaluated the effect of channel dimensions increment in the pressure drop and heat transfer rates and efficiency. The authors observed that for the same heat transfer rate and efficiency, the dimensions increment considerably reduced the pressure drop. The optimal operating conditions of the scaled-up heat exchanger were defined as the efficiency below 0.4, marked by a very reduced pressure drop. However, the authors concluded that the scale increment should be evaluated specifically according to the main application goal.

Mohammad et al., (2021), aiming the intensification of the Fischer-Tropsch process with microreactors, tested three scale-up arrangements: the use of parallel units (numbering-up), the use of units in series and the increment of channel dimensions (scale-out). The scaled-up channels consisted of 7 channels of 1000 μm x 1000 μm against the initial 11 channels of 500 μm x 500 μm. After the synthesis in the microreactors, the authors observed no significant effects of the increase in the channel width in the reactor performance.

A successful case of scale-up was presented by the company Ehrfeld Mikrotechnik BTS, which designed and manufactured a scaled-up millireactor with a production capacity of 10.000 ton/year. The MIPROWA millireactor, presented in Figure 6, was firstly designed in the micrometric scale. After the size scale increment, the millireactor presented a total width of 400 mm and length of 7 m with 150 rectangular channels with static mixers, achieving an operating condition of 1 m^3/h. The millireactor, when continuously operated, can replace up to 20 batch reactors. The investment in the millireactor development was motivated by the final product quality. The reaction yield was optimized, and the company achieved a quick return on the invested capital.

The scale-up can also be performed by the modular concept. Han et al., (2017), presented a multidimensional expansion strategy based on the modular integration, as illustrated in Figure 7. The throughput enlargement is carried out firstly by parallelizing N microchannels in a bidimensional matrixes. These M matrixes are then pilled to form a module. Finally, the integration of Q modules creates the system with a production flow rate proportional to the scale N x M x Q.

Adapted from Millireactor in Production (2016).

Figure 6. Milirreator MIPROWA.

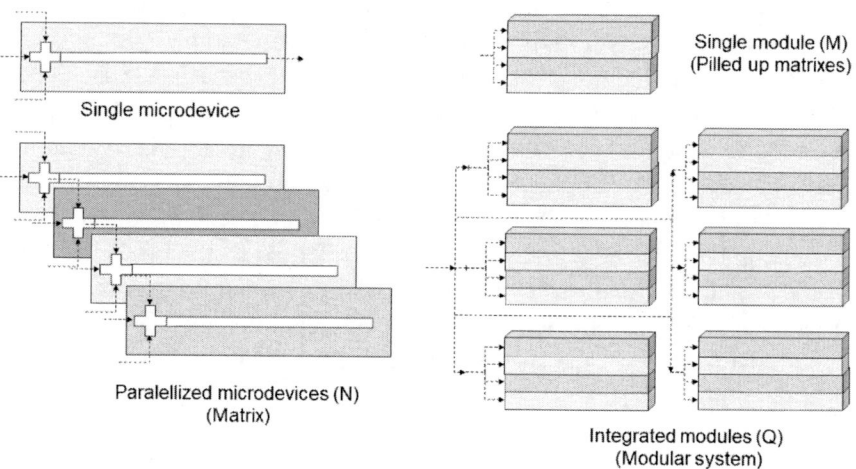

Adapted from Han et al., (2017) conceptualization.

Figure 7. Multidimensional scale-up strategy for modular plant assembly from a single device.

Tonkovich et al., (2005), developed a microchannel system for hydrogen generation on commercial scale. The system was composed by modules with

72 rectangular microchannels in parallel arrangement with dimensions of 4 mm of width and 0.5 mm of height. The modules with the grouped matrixes presented the dimensions of 45 cm (width) x 30 cm (length). The authors highlighted the importance of the uniform flow among the microdevices to ensure the efficient production. For the studied case, flow maldistribution below 20%, i.e., the mass flow rate ranging by about 20% at each reactor inlet, did not present significant effects in the system performance.

One of the major challenges is to ensure the uniform flow distribution in each of the microdevice units. The reactants maldistribution can reduce the microdevice performance, once the optimal reactant proportion could be incorrect, deprecating the reactor performance. Wang et al., (2016), approached the difficulty of keeping the transport characteristics, and reaction performance observed in a single microreactor when the numbering-up is performed. The authors highlighted the importance of an adequate design for the flow distributor. Accordingly, the flow distributors have the main goal of providing flow uniformity along all microdevices, contributing to keeping the high performance of the parallelized unit, similar to the observed in a single microreactor (Saber et al., 2010).

Shen et al., (2018), pointed out the importance of the uniform flow in several scaled-up systems, including single-phase and two-phase flows and multiphase flows with bubbles and droplets generation. For a specific situation, it is mandatory to evaluate the flow distribution to achieve an efficient scaled-up system. The studies about the flow distribution must be extended to contemplate a wider range of fluids, including viscous and non-Newtonian, for future applications in microchannels and scale-up procedures. Vladisavljević et al., (2013), reviewed industrial applications of lab-on-chip devices, presenting the parallelization as an efficient method to be applied in reactive systems. However, in this arrangement, the inlet flow must be equally distributed at all microdevice units. Although, the microchannels can present some imperfections at the surface walls, and the hydraulic resistance can be different among these channels, also contributing to flow maldistribution. In this context, the flow distribution is a key parameter related to the system efficiency. Dong et al., (2021), grouped the flow distributors in internal and external. An external flow distributor is an independent unit, where the reactants fluids are distributed before entering the microreactor channels. The internal distributor is manufactured coupled to the microchannels. In addition, these two approaches can be combined to increase the reaction performance.

A study of biodiesel production in a scaled-up microplant was performed by Lopes et al., (2019). Different conical flow distributors were evaluated.

After the initial tests, two distributors with flow deviations below 1% were selected to perform the ethanol and vegetable oil feed in the millidevices. Each reactant flow stream with its specific transport properties (dynamic viscosity) requires an optimal distributor design, aiming for the minimization of flow maldistribution. An integrated system, i.e., a micro-chemical plant, was evaluated. The ethanol flow deviation increased to 4.2% due to the pressure drop of the system. The scaled-up system was capable to achieve a biodiesel production rate of 126.4 mL min^{-1} with a yield of 83.3%.

Bannatham et al., (2021), carried out an experimental study of the transesterification in a microreactor with an inner diameter of 0.508 mm and length of 1.2 m. The reaction performance was evaluated at 52°C and 60°C for a methanol-oil molar ration of 6:1. The droplet-base microfluidics technique was used to feed the methanol in the continuous oil stream. The methanol droplet diameter was about 0.04 mm. The amount of methyl ester increased with temperature, flow rate, and reactor length. According to the authors, the increase in the reactor length caused the reduction of droplets sizes, resulting in an increment of the interfacial area and higher yield. In order to achieve a yield of over 97%, the microreactor length should be at least 1.2 m.

Lukić and Vrsaljko, (2020), developed millireactors to produce biodiesel. The conversion of sunflower oil in methyl esters of fatty acids was evaluated in channels with diameters from 1.5 mm to 3.5 mm. The millireactors of smaller diameters at higher temperatures achieved superior biodiesel yield with shorter times, regarding conventional batch reactors. As expected, the evaluated millireactor presented a larger production volume and smaller pressure drop concerning microreactors.

In this context, the study of scale-up strategies using micro/millireactors is of fundamental importance to enable its usage in an industrial scale. The microfluidic advantages are well known and established in the scientific community; however, there is still a necessity for studies focusing on the production throughput increment using microdevices and the particularities at each reactive process, including single and multiphase flows.

Conclusion and Future Perspectives

In the recent decades, microfluidic devices received great attention due to their inherent advantages of enhanced microscale transport phenomena is distinct technology fields, including chemical and biochemical process intensification. The need for renewable and sustainable energy sources on a global scale demands the development of innovative and efficient synthesis of biofuels. All these issues contributed to the development of microfluidic-based biodiesel processes. Major advances were achieved in fluid mixing and reaction performance, especially for biodiesel synthesis. However, there is still a lack in the effective and practical scale-up of the microfluidics devices to the industrial scale demand. The combination of different scale-up strategies, keeping the intensified heat and mass transfer rates, is a key-factor, since the process performance parameters, including selectivity and yield are directly affected by the transport phenomena. The coupling of scale-out and numbering-up concepts appears as an interesting alternative for a large throughput increment from a single optimized device, i.e., bringing together the best of both approaches. The scale-out from micro- to milli-scale, jointly with the use of optimal design static micromixers, allows the increment of operational flow rate, reducing the pressure drop and keeping the optimal mixing and reaction performance from the microscale. The numbering-up provides a prompt increase in the total throughput by a factor proportional to the microchemical plant construction, i.e., multiple by the number of parallelized microdevices, the pilled matrixes, and the integrated modules. Despite this smart and flexible strategy, a uniform flow distribution must be ensured along with all devices, minimizing variances among the microreactors. In all these steps, detailed CFD simulation can be employed for the devices design and optimization, appearing as an economical method, prior to the physical prototyping. Also, 3D printing can be used along CFD to manufacture microchemical plants. In this context, there are several advantages of the use of micro and millidevices in the continuous production of biodiesel, employing modern tools of Computer Aided Engineering and Industry 4.0. Future efforts must be spent in the scale-up and low-cost manufacturing strategies of integrated and automated modular plants, including mixing, reaction, and purification stages. These integrated plants will allow the continuous and flexible production of biodiesel, achieving the required demand.

References

Akkarawatkhoosith, N., Kaewchada, A. and Jaree, A. (2020). Continuous catalyst-free biodiesel synthesis from rice bran oil fatty acid distillate in a microreactor. *Energy Reports,* 545–549.

Akkarawatkhoosith, N., Tongtummachat, T., Kaewchada, A. and & Jaree, A. (2021). Non-catalytic and glycerol-free biodiesel production from rice bran oil fatty acid distillate in a microreactor. *Energy Conversion and Management: X,* 100096.

Anderson, J. D. J. (1995). *Computational Fluid Dynamics,* 1st ed. McGraw-Hill, New York.

ASTM, & D6751–15. (2015). D6751–15. *Standard, Specification for Biodiesel Fuel Blend Stock (B100) for Middle Distillate Fuels.* West Conshohocken, PA: ASTM.

Atadashi, I. M., Aroua, M. K., Aziz, A. R. A. and Sulaiman, N. M. N. (2012). High quality biodiesel obtained through membrane technology, *Journal of Membrane Science,* 154-164.

Bacic, M., Ljubic, A., Gojun, M., Salic, A., Tusek, A.J. and Zelic, B. (2021). Continuous integrated process of biodiesel production and purification—the end of the conventional two-stage batch process? *Energies,* 403.

Bakker, A, (2006). *Lecture 7 - Meshing.* Available at: http://www.bakker.org/dartmouth06/engs150/07-mesh.pdf. Accessed December 14, 2021.

Balbino, T. A., Serafin, J. M., Malfatti-Gasperini, A. A. and de La Torre, L. G. (2016). Microfluidic assembly of pDNA/cationic liposome lipoplexes with high pDNA loading for gene delivery. *Langmuir,* 1799–1807.

Bannatham, P., Banthaothook, C., Limtrakul, S., Vatanatham, T., Jaree, A., and & Ramachandran, P. A. (2021). Two-scale model for kinetics, design, and scale-up of biodiesel production. *Industrial & Engineering Chemistry Research,* 15972–15988.

Baroutian, S., Aroua, M. K., Aziz, A. R. A. and Sulaiman, N. M. N. (2012). TiO_2/Al_2O_3 membrane reactor equipped with a methanol recovery unit to produce palm oil biodiesel, *International Journal of Energy Research,* 120–129.

Baroutian, S., Aroua, M. K., Raman, A. A. A. and Sulaiman, N. M. N. (2011). A packed bed membrane reactor for production of biodiesel using activated carbon supported catalyst, *Bioresource Technology,* 1095–1102.

Bashir, M. A., Wu, S., Zhu, J., Krosuri, A., Khan, M. U., and Ndeddy Aka, R. J. (2022). Recent development of advanced processing technologies for biodiesel production: A critical review. *Fuel Processing Technology,* 227, 107120.

Basiri, M., Rahimi, M. and Mohammadi, H. B. (2016). Ultrasound-Assisted Biodiesel Production in Micro Reactors. *Iranian Journal of Chemistry and Chemical Engineering,* 22-32.

Bhatti, M. M., Marin, M., Zeeshan, A. and Abdelsalam, S.I. (2020). Editorial: Recent trends in computational fluid dynamics. *Frontiers in Physics,* 1–4.

Billo, R. E., Oliver, C. R., Charoenwat, R., Dennis, B. H., Wilson, P. A., Priest, J. W. and Beardsley, H. (2015). A cellular manufacturing process for a full-scale biodiesel microreactor. *Journal of Manufacturing Systems,* 409–416.

Bird, R. B., Stewart, W. E. and Lightfoot, E. N. (2006). *Transport Phenomena,* Revised 2nd Edition. Vol. 1, John Wiley Sons, Inc.

Bonet, J., Valentin, P., Ruiz, A. E. B., Petrica, I. and Llorens, J. (2014). Thermodynamic study of batch reactor biodiesel synthesis. *Revista de Chimie,* 358–361.

Bucciol, F., Colia, M., Calcio Gaudino, E. and & Cravotto, G. (2020). Enabling technologies and sustainable catalysis in biodiesel preparation. *Catalysts,* 988.

Buchori, L., Istadi, I. and Purwanto, P. (2016). Advanced chemical reactor technologies for biodiesel production from vegetable oils - A review. *Bulletin of Chemical Reaction Engineering, Catalysis,* 406–430.

Budžaki, S., Miljić, G., Tišma, M., Sundaram, S. and Hessel, V. (2017). Is there a future for enzymatic biodiesel industrial production in microreactors? *Applied Energy,* 124–134.

Cao, P. G., Tremblay, A. Y. and Dube, M. A. (2009). Kinetics of canola oil transesterification in a membrane reactor, *Industrial & Engineering Chemistry Research,* 2533-2541.

Chipurici, P., Vlaicu, A., Calinescu, I., Vinatoru, M., Vasilescu, M., Ignat, N. D. and Mason, T. J. (2019). Ultrasonic, hydrodynamic and microwave biodiesel synthesis – A comparative study for continuous process. *Ultrasonics Sonochemistry,* 38–47.

Chueluecha, N., Kaewchada, A. and & Jaree, A. (2017). Biodiesel synthesis using heterogeneous catalyst in a packed-microchannel. *Energy Conversion and Management,* 145–154.

Chung, T. J. (2002). *Computational Fluid Dynamics.* Cambridge University Press, Cambridge.

Crawford, E., Duquette, D., Grant, D., Gray, R., Musselman, B., Peacock, M. and Petersen, J. (2008). *Continuous Biodiesel Production with Continuous Liquid-Liquid Extraction and Online MS Analysis.* Available at: https://www.syrris.com/wp-content/uploads/2017/10/Continuous-Biodiesel-Production.pdf. Accessed January 12, 2022.

Demirbas, A., Demirba¸s, A. and Demirbas, D. (2008). Biodegradability of biodiesel and petrodiesel fuels. *Energy Sources, Part A: Recovery, Utilization, and Environmental Effects Biodegradability of Biodiesel and Petrodiesel Fuels,* 169–174.

Dimov, I. K., Garcia-Cordero, J. L., O'Grady, J. and O'Kennedy, R. (2008). Integrated microfluidic tmRNA purification and real-time NASBA device for molecular diagnostics. *Lab on a Chip,* 2071–2078.

Dong, Z., Wen, Z., Zhao, F., Kuhn, S. and Noël, T. (2021). Scale-up of micro- and milli-reactors: An overview of strategies, design principles and applications. *Chemical Engineering Science: X,* 100097.

Ebiura, T., Echizen, T., Ishikawa, A., Murai, K. and Baba, T. (2005). Selective transesterification of triolein with methanol to methyl oleate and glycerol using alumina loaded with alkali metal salt as a solid-base catalyst. *Applied Catalysis A: General,* 111–116.

Essamlali, Y., Amadine, O., Larzek, M., Len, C. and Zahouily, M. (2017). Sodium modified hydroxyapatite: Highly efficient and stable solid-base catalyst for biodiesel production. *Energy Conversion and Management,* 355–367.

Eze, V. C., Phan, A. N., Pirez, C., Harvey, A. P., Lee, A. F. and Wilson, K. (2013). Heterogeneous catalysis in an oscillatory baffled flow reactor. *Catalysis Science & Technology,* 2373.

Fereidooni, L., Abbaspourrad, A. and Enayati, M. (2021). Electrolytic transesterification of waste frying oil using Na^+/zeolite–chitosan biocomposite for biodiesel production. *Waste Management,* 48–62.

Fonseca, J. M., Teleken J. G., de Cinque Almeida V. and da Silva, C. (2019). Biodiesel from waste frying oils: methods of production and purification. *Energy Convers Manage*, 205-218.

Georgogianni, K. G., Katsoulidis, A. K., Pomonis, P. J., Manos, G. and Kontominas, M. G. (2009). Transesterification of rapeseed oil for the production of biodiesel using homogeneous and heterogeneous catalysis. *Fuel Processing Technology*, 1016–1022.

Gojun, M., Šalić, A. and Zelić, B. (2021). Integrated microsystems for lipase-catalyzed biodiesel production and glycerol removal by extraction or ultrafiltration. *Renewable Energy*, 213–221.

Gojun, M., Pustahija, L., Tušek, A. J., Šalić, A., Valinger, D. and Zelić, B. (2019). Kinetic parameter estimation and mathematical modelling of lipase catalysed biodiesel synthesis in a microreactor. *Micromachines*, 759.

Gopi, R., Thangarasu, V., Vinayakaselvi M, A. and Ramanathan, A. (2022). A critical review of recent advancements in continuous flow reactors and prominent integrated microreactors for biodiesel production. *Renewable and Sustainable Energy Reviews*, 111869.

Guan, G. and Kusakabe, K. (2009). Synthesis of biodiesel fuel using an electrolysis method. *Chemical Engineering Journal*, 159–163.

Guan, G., Kusakabe, K. and Yamasaki, S. (2009). Tri-potassium phosphate as a solid catalyst for biodiesel production from waste cooking oil. *Fuel Processing Technology*, 520–524.

Guan, G., Teshima, M., Sato, C., Son, S. M., Irfan, M. F., Kusakabe, K., Ikeda, N. and Lin, T.-J. (2010). Two-phase flow behavior in microtube reactors during biodiesel production from waste cooking oil. *AIChE Journal*, 1383–1390.

Gupta, J., Agarwal, M. and Dalai, A. K. (2019). Intensified transesterification of mixture of edible and non-edible oils in reverse flow helical coil reactor for biodiesel production. *Renewable Energy*, 509–525.

Han, W., Charoenwat, R. and Dennis, B. H. (2011). Numerical investigation of biodiesel production in capillary microreactor, in: Proceedings of the ASME Design Engineering Technical Conference. *American Society of Mechanical Engineers Digital Collection*, 253–258.

Han, T., Zhang, L., Xu, H. and Xuan, J. (2017). Factory-on-chip: Modularised microfluidic reactors for continuous mass production of functional materials. *Chemical Engineering Journal*, 765-773.

Hirt, C. W. and Nichols, B. D. (1981). Volume of Fluid (VOF) method for the dynamics of free boundaries. *Journal of Computational Physics*, 201-225.

Jachuck, R., Pherwani, G., & Gorton, S. M. (2009). Green engineering: continuous production of biodiesel using an alkaline catalyst in an intensified narrow channel reactor. *Journal of Environmental Monitoring*, 642-647.

Kang, S. W. and Tseng, S. C. (2007). Analysis of effectiveness and pressure drop in microcross-flow heat exchanger. *Applied Thermal Engineering*, 877–885.

Kefas, H. M., Yunus, R., Rashid, U. and Taufiq-Yap, Y. H. (2019). Enhanced biodiesel synthesis from palm fatty acid distillate and modified sulfonated glucose catalyst via an oscillation flow reactor system. *Journal of Environmental Chemical Engineering*, 102993.

Kiss, A. A. and Bildea, C. S. (2012). A review of biodiesel production by integrated reactive separation technologies, *Journal of Chemical Technology & Biotechnology*. 861-879.

Kobayashi, I., Neves, M. A., Yokota, T., Uemura, K. and Nakajima, M. (2009). Generation of geometrically confined droplets using microchannel arrays: effects of channel and step structure. *Industrial & Engineering Chemistry Research*, 8848–8855.

Kralj, J. G., Sahoo, H. R. and Jensen, K. F. 2007. Integrated continuous microfluidic liquid–liquid extraction. *Lab Chip*, 256-263.

Kusdiana, D. and Saka, S. (2001). Kinetics of transesterification in rapeseed oil to biodiesel fuel as treated in supercritical methanol. *Fuel*, 693–698.

Laziz, M., KuShaari, A., Azeem, K. Z., Yusup, B., Chin, S. and Denecke, J. (2020). Rapid production of biodiesel in a microchannel reactor at room temperature by enhancement of mixing behaviour in methanol phase using volume of fluid model. *Chemical Engineering Science*, 115532.

Lin, C. Y., Chen, Y. Y., Chen, P. Y., Chen, M. C., Su, T. F. and Chiang, Y. Y. (2021). Scale-out production in core-annular liquid–liquid microextractor. *Journal of Flow Chemistry*, 569–577.

Lomax, H., Pulliam, T. H. and Zingg, D. W. (1999). *Fundamentals of Computational Fluid Dynamics*. University of Toronto Institute for Aerospace Studies.

Lopes, M. G. M., Santana, H. S., Andolphato, V. F., Russo, F. N., Silva Jr., J. L. and Taranto, O. P. (2019). 3D printed micro-chemical plant for biodiesel synthesis in millireactors. *Energy Conversion and Management*, 475-487.

López-Guajardo, E., Ortiz-Nadal, E., Montesinos-Castellanos, A. and Nigam, K. D. P. (2017). Process intensification of biodiesel production using a Tubular Micro-Reactor (TMR): Experimental and numerical assessment. *Chemical Engineering Communications*. 467–475.

Lukić, M. and Vrsaljko, D. (2021). Effect of channel dimension on biodiesel yield in millireactors produced by stereolithography. *International Journal of Green Energy*, 156–165.

Maia, D. C., Salim, V. M. M. and Borges, C. P. (2016). Membrane contactor reactor for transesterification of triglycerides heterogeneously catalyzed, *Chemical Engineering and Processing*, 220-225.

Martínez Arias, E. L., Fazzio Martins, P., Jardini Munhoz, A. L., Gutierrez-Rivera, L. and Maciel Filho, R. (2012). Continuous synthesis and in situ monitoring of biodiesel production in different microfluidic devices. *Industrial & Engineering Chemistry Research*, 10755–10767.

Mathes, B. (2016). *Future Production Concepts in the Chemical Industry*. Available at: https://www.chemanager-online.com/en/topics/production/future-production-concepts-chemical-industr. Accessed: January 16, 2022.

Mohadesi, M., Aghel, B., Maleki, M. and Ansari, A. (2020a). Study of the transesterification of waste cooking oil for the production of biodiesel in a microreactor pilot: The effect of acetone as the co-solvent. *Fuel*, 117736.

Mohadesi, M., Aghel, B., Maleki, M. and Ansari, A. (2020b). The use of KOH/Clinoptilolite catalyst in pilot of microreactor for biodiesel production from waste cooking oil. *Fuel*, 116659.

Mohadesi, M., Gouran, A. and Dehghan Dehnavi, A. (2021). Biodiesel production using low cost material as high effective catalyst in a microreactor. *Energy*, 119671.

Mohammad, N., Chukwudoro, C., Bepari, S. And Basha, O., Aravamudhan, S. and Kuila, D. (2021). Scale-up of high-pressure F-T synthesis in 3D printed stainless steel microchannel microreactors: Experiments and modeling. *Catalysis Today*.

Moukalled, F., Mangani, L. and Darwish, M., (2016). *The Finite Volume Method in Computational Fluid Dynamics: An Advanced Introduction with OpenFOAM and Matlab, Fluid Mechanics and its Applications*. 1 edition, Vol. 1, Springer, Switzerland.

Nasiri, R., Shamloo, A., Akbari, J., Tebon, P., R. Dokmeci, M. and Ahadian, S. (2020). Design and simulation of an integrated centrifugal microfluidic device for CTCs separation and cell lysis. *Micromachines*, 699.

Ogunkunle, O. and Ahmed, N. A. (2019). A review of global current scenario of biodiesel adoption and combustion in vehicular diesel engines. *Energy Reports*, 1560–1579.

Palm, O. M., Barbosa, S. L. A. F., Gonçalves, M. W., Duarte, D. A., Catapan, R. C. and Pinto, C. R. S. C. (2022). Plasma-assisted catalytic route for transesterification reactions at room temperature. *Fuel*, 121740.

Pavlović, S., Šelo, G., Marinković, D., Planinić, M., Tišma, M. and Stanković, M. (2021). Transesterification of sunflower oil over waste chicken eggshell-based catalyst in a microreactor: an optimization study. *Micromachines*, 120.

Qadeer, M. U., Ayoub, M., Komiyama, M., Daulatzai, M. U. K., Mukhtar, A., Saqib, S., Ullah, S., Qyyum, M. A., Asif, S. and Bokhari, A. (2021). Review of biodiesel synthesis technologies, current trends, yield influencing factors and economical analysis of supercritical process. *Journal of Cleaner Production*, 127388.

Quiroz-Pérez, E., Gutiérrez-Antonio, C. and Vázquez-Román, R. (2019). Modelling of production processes for liquid biofuels through CFD: A review of conventional and intensified technologies. *Chemical Engineering and Processing: Process Intensification*, 107629.

Rahimi, M., Aghel, B., Alitabar, M., Sepahvand, A. and Ghasempour, H. R. (2014). Optimization of biodiesel production from soybean oil in a microreactor. *Energy Conversion and Management*, 599–605.

Rahimi, M., Mohammadi, F., Basiri, M., Parsamoghadam, M. A. and Masahi, M. M. (2016). Transesterification of soybean oil in four-way micromixers for biodiesel production using a cosolvent. *Journal of the Taiwan Institute of Chemical Engineers*, 203–210.

Reyes, I., Ciudad, G., Misra, M., Mohanty, A., Jeison, D. and Navia, R. (2012). Novel sequential batch membrane reactor to increase fatty acid methyl esters quality at low methanol to oil molar ratio, *Chemical Engineering Journal*, 459-467.

Rossetti, I. (2018). Continuous flow (micro-)reactors for heterogeneously catalyzed reactions: Main design and modelling issues. *Catalysis Today*, 20–31.

Saber, M., Commenge, J. M. and Laurent, F. (2010). Microreactor numbering-up in multi-scale networks for industrial-scale applications: Impact of flow maldistribution on the reactor performances. *Chemical Engineering Science*, 372-379.

Saka, S., and Kusdiana, D. (2001). Biodiesel fuel from rapeseed oil as prepared in supercritical methanol. *Fuel*, 225–231.

Santana, H. S., Lopes, M. G. M., Silva Jr., J. L. and Taranto, O. P. (2018a). Application of microfluidics in process intensification. *International Journal of Chemical Reactor Engineering*, 16(12).

Santana, H. S., Sanchez, G. B. and Taranto, O. P. (2017a). Evaporation of excess alcohol in biodiesel in a microchannel heat exchanger with Peltier module. *Chemical Engineering Research and Design*, 20-28.

Santana, H. S., Silva, J. L., Da Silva, A. G. P., Rodrigues, A. C., Amaral, R. D. L., Noriler, D. and Taranto, O. P. (2021). Development of a new micromixer "elis" for fluid mixing and organic reactions in millidevices. *Industrial & Engineering Chemistry Research*, 9216–9230.

Santana, H. S., Silva, J. L., Tortola, D. S., & Taranto, O. P. (2018b). Transesterification of sunflower oil in microchannels with circular obstructions. *Chinese Journal of Chemical Engineering*, 852–863.

Santana, H. S., Tortola, D. S., Reis, É. M., Silva, J. L. and Taranto, O. P. (2016). Transesterification reaction of sunflower oil and ethanol for biodiesel synthesis in microchannel reactor: Experimental and simulation studies. *Chemical Engineering Journal*, 752–762.

Santana, H. S., Tortola, D. S., Silva, J. L. and Taranto, O. P. (2017b). Biodiesel synthesis in micromixer with static elements. *Energy Conversion and Management*, 28–39.

Silva, M. V. D., Hori, C. E. and Reis, M. H. M. (2015). Thermochemical data of the oleic acid esterification reaction: A quantum mechanics approach. *Fluid Phase Equilibria*, 168–174.

Shen, Q., Zhang, C., Tahir, M. F., Jiang, S., Zhu, C., Ma, Y. and Fu, T. (2018). Numbering-up strategies of micro-chemical process: Uniformity of distribution of multiphase flow in parallel microchannels. *Chemical Engineering & Processing: Process Intensification*, 148-159.

Shuit, S. H., Ong, Y. T., Lee, K. T., Subhash, B. and Tan, S. H. (2012). Membrane technology as a promising alternative in biodiesel production: a review, *Biotechnology Advances*, 1364-1380.

Stacy, C. J., Melick, C. A. and Cairncross, R. A. (2014). Esterification of free fatty acids to fatty acid alkyl esters in a bubble column reactor for use as biodiesel. *Fuel Processing Technology*, 70–77.

Sun, J., Ju, J., Ji, L., Zhang, L., and Xu, N. (2008). Synthesis of biodiesel in capillary microreactors. *Industrial & Engineering Chemistry Research*, 1398–1403.

Tabatabaei, M., Aghbashlo, M., Dehhaghi, M., Panahi, H. S. P., Mollahosseini, A., Hosseini, M. and Soufiyan, M. M. (2019). Reactor technologies for biodiesel production and processing: A review. *Progress in Energy and Combustion Science*, 239–303.

Thangarasu, V., Siddharth, R. and Ramanathan, A. (2020). Modeling of process intensification of biodiesel production from Aegle Marmelos Correa seed oil using microreactor assisted with ultrasonic mixing. *Ultrasonics Sonochemistry*, 104764.

Tiwari, A., Rajesh, V. M. and Yadav, S. (2018). Biodiesel production in micro-reactors: a review. *Energy for Sustainable Development*, 143-161.

Tonkovich, A., Kuhlmann, D., Rogers, A., Mcdaniel, J., Fitzgerald, S., Arora, R. and Yuschak, T. (2005). Microchannel technology scale-up to commercial capacity. *Chemical Engineering Research and Design*, 634-639.

Vankayala, B. K., Lob, P., Hessel, V., Menges, G., Hofmann, C., Metzke, D., Krtschil, U. and Kost, H. J. (2007). Scale-up of process intensifying falling film microreactors to pilot production scale. *International Journal of Chemical Reactor Engineering*, A91.

Vladisavljević, G. T., Khalid, N., Neves, M. A., Kuroiwa, T., Nakajima, M., Uemura, K., Ichikawa, S. and Kobayashi, I. (2013). Industrial lab-on-a-chip: Design, applications and scale-up for drug discovery and delivery. *Advanced Drug Delivery Reviews*, 1626-1663.

Voegele, E. (2020). NBB: Biodiesel can help meet USDA's AIA goals. *Biodiesel Magazine*. Available at: http://www.biodieselmagazine.com/articles/2517256/nbb-biodiesel-can-help-meet-usdaundefineds-aia-goals. Accessed October 15, 2021.

Wang, L., Kong, X. and Qi, Y. (2016). Optimal design for split-and-recombine-type flow distributors of microreactors based on blockage detection. *Chinese Journal of Chemical Engineering*, 897-903.

Wen, Z., Yu, X., Tu, S. T., Yan, J. and Dahlquist, E. (2009). Intensification of biodiesel synthesis using zigzag micro-channel reactors. *Bioresource Technology*, 3054–3060.

Wirth, T. (2013). *Microreactors in Organic Chemistry and Catalysis*. (T. Wirth, Ed.), 2nd edition, Vol. 1, Wiley-VCH.

Whitesides, G. M. (2006). The origins and the future of microfluidics. *Nature*, 368-373.

Xie, W., Peng, H. and Chen, L. (2006). Transesterification of soybean oil catalyzed by potassium loaded on alumina as a solid-base catalyst. *Applied Catalysis A: General*, 67–74.

Xu, W., Gao, L. J., Xiao, G. M. (2015). Biodiesel production optimization using monolithic catalyst in a fixed-bed membrane reactor, *Fuel*, 484-490.

Yeh, S. I., Huang, Y. C., Cheng, C. H., Cheng, C. M. and Yang, J. T. (2016). Development of a millimetrically scaled biodiesel transesterification device that relies on droplet-based co-axial fluidics. *Scientific Reports*, 29288.

Zabeti, M., Wan Daud, W. M. A. and Aroua, M. K. (2009). Activity of solid catalysts for biodiesel production: A review. *Fuel Processing Technology*, 770–777.

Zhang, B., Ren, J., Liu, X., Guo, Y., Guo, Y., Lu, G. and Wang, Y. (2010). Novel sulfonated carbonaceous materials from p-toluenesulfonic acid/glucose as a high-performance solid-acid catalyst. *Catalysis Communications*, 629–632.

Zhang, J., Wang, K., Teixeira, A. R., Jensen, K. F. and Luo, G. (2017). Design and scaling up of microchemical systems: A review. *The Annual Review of Chemical and Biomolecular Engineering*, 285-305.

Zhang, X., Wiles, C., Painter, S. L., Watts, P. and Haswell, S. J. (2006). Microreactors as tools for chemical research. *Chimica Oggi*, 43–45.

Chapter 3

Effect of Biodiesels – Bioethanol Fuel Mixture on Performance, Characteristics in Diesel Engines

M. Acaroğlu[1,*], PhD and H. Köse[2,†], PhD

[1]Department of Mechanical Engineering,
Selçuk University, Konya, Turkey
[2]Department of Automotive Technology,
Ege University, İzmir, Turkey

Abstract

Cynara cardunculus is a Mediterranean perennial herb. Cynara cardunculus is a herbaceous, robust, long-lived plant that can be grown even in non-agricultural lands in Turkey, with low water requirement, high yield. In this study, the effects of Cynara biodiesel-bioethanol-diesel fuel mixtures on engine performance characteristics (engine power, torque, emissions) were investigated. Cynara cardunculus biodiesel was mixed with different proportions of bioethanol diesel fuel. It has been named as diesel (D100), biodiesel (B100) and diesel-biodiesel-bioethanol (B5E5-B7E5).

According to the experimental results, maximum engine power (44.44 kW) and maximum torque (166 Nm) B7E5, the highest thermal efficiency was acquired with B7E5 fuel mixture at 3000 rpm. The lowest specific fuel consumption (be) was measured with D100 fuel at 2000 rpm as 218.18 g/kWh. Maximum in-cylinder pressure was measured as 103.2 bar B5E5 at 3000 rpm. Max heat release rates (HRR) were obtained with B7E5 at 2000 rpm and 3000 rpm as 524 kJ/m3.deg and 602.4 kJ/m3.deg, respectively. CO emission values for B100, B5E5 and B7E5 fuel mixtures decreased averagely

---

[*] Corresponding Author's Email: acaroglu@selcuk.edu.tr
[†] Corresponding Author's Email: huseyin.kose@ege.edu.tr

In: The Future of Biodiesel
Editor: Michael F. Simpson
ISBN: 979-8-88697-166-8
© 2022 Nova Science Publishers, Inc.

82.84%, 42.975% and 52.73% compared to standard diesel operation. CO_2 emissions increased averagely 2.1%, 1.8%, 3.04%. HC emissions significantly reduced. NOx emissions for all fuel increased. Consequently, cynara cardunculus seed oil biodiesel and bioethanol can be mixed at certain rates at diesel engine.

Keywords: cynara cardunculus, cylinder pressure, bioethanol, emission, biodiesel, HRR

Nomenclature

B100	Biodiesel 100%
B5E5	Blend of biodiesel 5% and bioethanol 5%
B7E5	Blend of biodiesel 7% and bioethanol 5%
BTE	Brake thermal efficiency (%)
BSFC	Brake specific fuel consumption (kg/kWh)
BD	Biodiesel
CI	Compression ignition
CO	Carbon monoxide (%)
CO_2	Carbon dioxide (%)
Diesohol	Diesel-Ethanol
D100	Diesel 100%
HRR	Heat release rate (kJ/m^3 deg)
HC	Hydrocarbon (ppm)
THC	Total Hydrocarbon (ppm)
NO_X	Nitrogen oxides (ppm)
RPM	Revolution per minute

Introduction

Diesel engines (CI) are used in different fields such as agriculture, transportation, industry, commercial areas due to their low fuel consumption and high efficiency with high engine power output. However, the rapid depletion of fossil fuels used in these vehicles, increased prices, the cause of both environmental pollution and climate change and most importantly, stricter emission norms has encouraged the exploration and use of alternative fuels. In this context, biodiesel and bioalcohol mixtures as alternative fuels are the most used in diesel engines due to, they reduce emissions such as smoke, HC, and CO during the combustion process (Köse and Ciniviz 2013, Sukjit, Herreros et al. 2013). Biodiesel and bioethanol are renewable fuels obtained from vegetable feedstocks or animal oils. These oils are extremely important to meet the fuel demand resulting from the depletion of fossil fuels in the future. The biodiesel obtained from these oils can be used partially or completely as fuel in the CI engine.

In this context, Cynara cardunculus, which is among the non- food plants, the most important of the multi annual plants, is excellent properties in biofuel production can be grown in low priced arid regions and need less irrigation and fertilization.

The biggest advantage of Cynara cardunculus compared to other plants is the low irrigation requirement and growing cost. Another advantage is that it is a potential product for producing low energy cost biodiesel. It has approximately 61% linoleic, 23.5% oleic, 12.0% palmitic and 3.5% stearic acid fat formation. It has a structure like the physicochemical properties of sunflower oil with more than 25% oil content in the seeds. It is possible to obtain Cynara cardunculus seed oil 400-650 kg/ha.year. With the oil yield of its seeds, it has a structure like that of sunflower and soybean. The lower heating value of Cynara cardunculus seed oil is 41 MJ/kg and the cetane number is 51.

It is very important to produce biodiesel from the plant that it has a high oil content and high lower heat value. Due to its high cetane number (CN), biodiesel can be used as a direct fuel or mixed with diesel fuel in direct injection or normal diesel engines. Moreover, the plant can be used in different sectors (in paper and pulp production, chemical industry, animal feed industry, pharmacologic active compounds etc.), has increased the importance of this plant in recent years. Also, it is a multi-year resistant plant that will grow in arid regions that do not need irrigation and using it in these areas, biodiesel production costs will be reduced significantly (Piscioneri, Sharma et al. 2000, Curt, Sanchez et al. 2002, Pasqualino 2006, Rakopoulos 2012).

The advantages of biodiesel fuel as diesel fuel are that it has minimum sulfur content, high flash point, high oxygen content, high CN for a better ignition and high lubricating properties (Jamrozik, Tutak et al. 2017). Another advantage is that since biodiesel is nonaromatic which decreases particulate matter formation. On the other hand, the major disadvantages of biodiesel are its high viscosity, high yield point, low heating value and the increase it causes in NOx emissions (Aydin and İlkılıç 2010). One way to solve these problems is to mix alcohols into diesel fuel. Among these alcohols, ethanol and methanol are used the most. Ethanol, especially the raw material of alcoholic beverages and an organic compound. Ethanol is included in all vehicles from the factory exit. Ethanol is a kind of fuel and biofuel (Ozcelik, Acaroglu et al. 2018, Teoh, Yu et al. 2019).

It has high octane number, wide flammability limit, high burning rate high evaporation temperature and high oxygen content of 35%. On the other hand, ethanol cannot be used in CI engines directly due to the properties of ethanol such as oxidation stability, insoluble in diesel fuel, low calorific value, and flash point. To overcome these technical obstacles, ethanol can be confused with biodiesel, which is expected to improve these properties. In addition, biodiesel allows ethanol to dissolve in diesel fuel. This causes triple fuel mixtures (diesel-bioethhanol-biodiesel) to become stable at zero degrees Celsius. Meanwhile, NOx and PM emissions can be reduced at the same time if the combustion temperature can be lowered by adding ethanol (Micic and Jotanovic 2015, Nantha Gopal, Ashok et al. 2017).

There is many research which have been made to provide the solubility of diesel-ethanol mixtures with different additives (emulsifiers), but additives only contribute to solubility and cause the amount of ethanol in the mixture to decrease (Pidol, Lecointe et al. 2012, An, Yang et al. 2015). Moreover, studies indicated that the emulsifiers used to mix diesel and ethanol had no effect on the falling flash point. However, biodiesel produced from different raw materials is reported that functions as an emulsifier for ethanol due to provides more dissolution of ethanol in diesel fuel prevents phase decomposition and increases the cetane number. In addition to all these advantages, the use of biodiesel instead of emulsifier in diesel fuels increases the flash point, which is an important parameter for the storage of the mixture, improves the viscosity and density of the mixture, brings it closer to diesel fuel, and increases the lubricant feature, which is also important for the fuel system. Thus, this disadvantage of can be eliminated with biodiesel added to diesel-ethanol fuel mixtures and used technically without major changes in existing diesel engine (G. Venkata Subbaiah, et al. 2010, Hulwan and Joshi 2011, Yilmaz, Vigil et al. 2014, Tse, Leung et al. 2015, Parthasarathy, Isaac JoshuaRamesh Lalvani et al. 2016). Although many studies have been carried out on biodiesel, ethanol, diesel, mixtures, the use of triple mixtures in the literature is relatively limited. Therefore, the use of biodiesel-ethanol in combination with diesel can result in a significant reduction of pollutant emissions. Zhu, Cheung et al. (2010) performed an experimental study with diesel engine using 5%, 10%, 15% ethanol or methanol and pure biodiesel fuel blends at a constant speed 1800 rpm.

They conducted research to reduce NOx emissions and particulate matter using fuel blends. They found that as the amount of ethanol in the mixture increased, NOx emissions and particulate emissions decreased significantly, while HC and CO emissions increased. Armas, Martínez-Martínez et al. (2011) comparative experimental study was conducted to determine the effect of fuel mixtures on common rail system components used in diesel engine. They were used to be tested two fuels using two Bosch fuel injection systems consisting of common rail system and piezoelectric fuel injector. The first of the systems tests o were carried out using low sulfur diesel fuel. The second tests were carried out with a ternary blend (ethanol (7.7%)-biodiesel (27.69%)-diesel (69.61%)).

Injection systems were operated under at 2500 rpm and 1500 bar at injection pressure, during 600 hours for 12 hours/day. They found that the ternary blends showed similar effect on system components when compared to diesel fuel. Lee, Liu et al. (2011) experimental study was conducted to determine the effects on energy efficiencies and emissions adding ethanol (4%) -butanol (1%)-diesel (65-90%)-biodiesel (5-30%) fuel blends using in a diesel generator. They found that the fuel blends gave stable mixtures in 30 days. Specific fuel consumption increased 0.45 to 1.6% for the BD1041 fuel mixture, NOx emission decreased by 2.8-6.0%, particulate matter by 12.6-23.7% and total hydrocarbons by 20.4-23.8% as compared to conventional diesel. de Oliveira, Valente et al. (2018) experimental study was conducted to on 44 kW diesel engine at 1800 rpm and a constant load of 27.4 kW/min and adding 5%, 10%, 15% and 20% ethanol and 7% and 20% biodiesel into the diesel fuel. They found that the B7E5 biodiesel- ethanol blend reduced CO emissions up to 7% and THC emissions with ethanol addition (B7E15) by 14%. On the other hand, CO_2 and THC emissions increased with the rate of biodiesel increasing from 7% to 20%. Kwanchareon, Luengnaruemitchai et al. (2007) examined the fuel stability and physical properties of ternary blends by using different ethanol rate and temperatures. They found that the physical properties of the fuel mixture, which contains 5% ethanol, were like diesel fuel, but found that the flash point decreased. While CO and HC emissions decreased, NOx emissions increased as engine load increased. Yilmaz (2012) biodiesel-methanol-diesel (BMD) (45-10-45%), biodiesel-ethanol-diesel (BED) (40%-20%-40%), BED (45% - 10% -%) 45) and BED (40% - 20% - 40%) fuel blends used in a CI engine.

They found that biodiesel-ethanol-diesel (BED) blends showed higher specific fuel consumption, CO with HC emissions and lower NOx emissions. In addition, while CO and HC emissions are reduced for methanol mixtures, ethanol mixtures are effective in reducing NOx emissions. Aydoğan (2015) experimental study was carried out for determining the effect of using E20B20D60, E30B20D50, E50B20D30 fuel mixtures on performance and emissions values in a diesel engine. He found that the values of engine power and torque with CO, HC and smoke emissions decreased, while NOx and specific fuel consumption (be) values increased with the use of ternary fuel blends. In the present study, cynara biodiesel-bioethanol- diesel fuel blends are considered as alternative fuel for use in CI engine. Main aim of this study investigates the effect on all engine characteristics of addition of cynara biodiesel (B100), diesel (D100) and biodiesel-bioethanol-diesel (B5E5-B7E5) blends compared to diesel fuel. Bioethanol was obtained from sugar beet plant. Characteristics of Cynara cardunculus and fuel mixtures are presented in Table 1. There was little works related to the use of cynara cardunculus seed oil biodiesel in ternary blends for DI engines. It is also hoped that the production of biofuels from regional raw materials will contribute to agriculture and reduce the imported oil.

Table 1. Characteristics of cynara and fuel blends

Properties	Unit	Bioethanol	Diesel	Cynara	B5E5	B7E5
Molecular weight	kg/mol	46.07	205	292	-	-
Carbon content	%	52	87	71	-	-
Hydrogen content	%	13	13	12	-	-
Oxygen content	%	34.8	0	10.8		
Density at 15°C (EN12185)	kg/m^3	794	845	876	833	834
Freezing point	°C	-114	-30--15	-4	-	-
Latent heating	kJ/kg	923	233	-	-	-
Lower heating value	MJ/kg	26.8	46.4	41.6	45.67	45.72
Octane number	-	108	-	-	-	-
Cetane index	-	-	50	51	48	49
Flash point	°C	12.8	>55°C	170-175	-	-
Viscosity *1000	Ns/m^2	1.5	2.5-3.5	4.975	2.63	2.83

Method

Experimental Test

The experiments were carried out on a 4-cylinder, 4-stroke, direct injection, turbocharged compression ignition (CI) 1.9 L engine. The general characteristics of the engine used in the experiments are presented in Table 2.

Table 2. Technical specifications of the engine

Engine (1.9 Multijet)	Technical specification
Cylinder number	4
Total cylinder volume	1910 JTD (cc)
Compression value (ratio)	18:1
Max Power kW – (rpm)	77 – (4000)
Max Torque Nm – (rpm)	200 – (1750)
Fuel System properties	Diesel Common Rail
Turbine	Turbo and intercooler
Diameter -Stroke (mm)	82 x 90.4

Table 3. Dynamometer and the load cell properties

Hydraulic dynamometer (BT190)	Technical properties
Max brake power	160 HP
Max brake torque	750 Nm
Max speed	6000 rpm
Water capacity	2.4 m3
Load cell capacity	2000 N (0.1%)

A TEDEA 3410 type load cell was used to determine the dynamometer load. The hydraulic dynamometer and the load cell specifications are presented Table 3.

In the trials, the engine speed was measured with the speed sensor coupled to the dynamometer. Engine in-cylinder pressure was determined using an AVL sensor and crank angle was measured using an AVL 365C crank angle encoder. Pressure values were measured and recorded at every 0.5 degrees of the crankshaft during 120 revolutions, and the average values were calculated. All these values were plotted with the AVL software CONCERTO. The experimental test setup is shown in Figure 1.

Exhaust emission values and temperatures of the engine were measured with K type thermocouple and Bosch BEA exhaust emission device. General specifications of Bosch BEA device and thermocouple are presented in Table 4.

Trials were made according to Turkish Standards 1231 (TS-1231). Unrefined oil was obtained by cold pressing under high pressure. The acid value of the unrefined Cynara cardunculus oil was 7.79 mg NaOH/g and the iodine value was 118.43. The catalysts used to reduce the acid value and iodine number and apply the transesterification method are NaOH and H_3PO_4 with an oil concentration of 2% by weight, respectively. Cynara was first heated to 70°C and stirred at 600 rpm with a magnetic stirrer. The alcohol-oil molar ratio was determined as 6:1, the fatty acid value was reduced below 1.0 mg KOH/g and the excess methanol, sulfuric acid and NaOH solution formed during the esterification reaction were removed from the surface. Samples were washed to completely remove methanol and catalyst. The acid value of the oil phase was determined by the EN14214 standard test method.

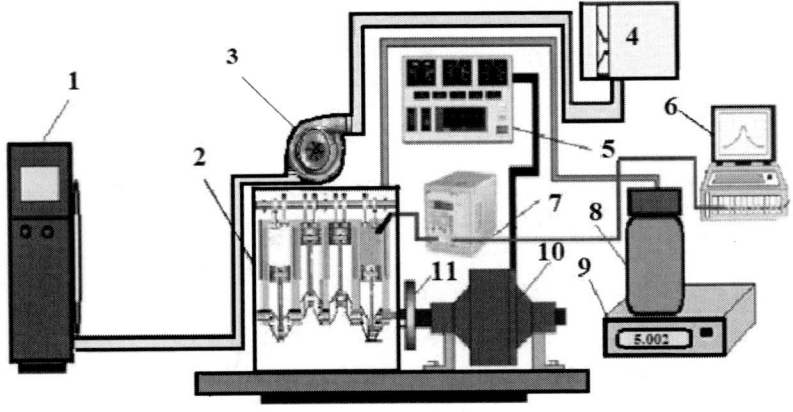

1. Exhaust Emissions Device
2. Engine
3. Turbocharger
4. Orifice Plate
5. Control Panel
6. AVL cylinder pressure meter
7. Charge Amplificatory
8. Fuel Tank
9. Digital Scales
10. Hydraulic Dynameters
11. Clutch

Figure 1. Experimental test assembly.

Table 4. Technical specifications Bosch BEA exhaust emission device and thermocouple

Bosch BEA 350	Measure range	Precision
CO	0,000...10.00	% 0.001 vol
CO_2	0,00...18.00	% 0.01 vol
HC	0... 9999	1 ppm vol
NOx	0..,5000	1 ppm vol
O_2	0.00 ... 22.00	% 0.01 vol
Lamda	0.500 ... 9.999	0.001
K-type thermocouple	0 ....1200 °C	±2 °C

Experimental Procedure

The hydraulic dynamometer used in the experiments was connected to engine by a shaft to ensure the engine load. The engine was operated to bring it up to standard operating temperature and the experiments were started. All tests were carried out at full throttle. Afterwards, the experiments were started. All fuel mixes during the experiment were recorded engine performance and emission values. To determine standard engine values, the engine was operated using diesel (D100) as control fuel. Then 5% biodiesel- 5% bioethanol- 90% diesel (B5E5), 7% biodiesel- 5% bioethanol- 90% diesel (B7E5) and finally 100% biodiesel (B100) fuels were tested. The experiments were carried out at 7 different engine speeds (between 1000 and 4000 rpm with an increase of 500 rpm) at full load. After each test, the fuel filter was changed, and the engine was run without load for approximately 30 minutes to remove the mixture in the fuel system.

Procedure Analyze

The following equations were used for the parameters measured in the tests performed with fuel mixtures:
Engine power,

$$P = [(F.L.n/9549)](kW) \qquad (1)$$

Brake thermal efficiency (for diesel fuel),

$$\eta = [(P/\dot{m}_D.LHV_D).100] \qquad (2)$$

Brake thermal efficiency for ternary mixture,

$$\eta = [(P/(\dot{m}_D.LHV_D + \dot{m}_{BD} LHV_{BD} + \dot{m}_E LHV_E)).100] \qquad (3)$$

Specific fuel consumption,

$$bsfc = [(\dot{m}_D + \dot{m}_{BD} + \dot{m}_E/P).3600](g/kW.h) \qquad (4)$$

The rate of HRR is found with the help of equation 5 by using the first rule of thermodynamics. This model most widely used one in the literature was developed by Krieger and Borman (Buyukkaya 2010, Fathi, Khoshbakhti Saray et al. 2010).

$$\dot{Q} = \frac{1}{\gamma - 1}\left[\gamma \cdot P \frac{dV}{d\theta} + V \frac{dP}{d\theta}\right] \tag{5}$$

γ = Specific heat ratio
θ = Crank angle

Using the equations 6-7-8 given below, the amount of instantaneous heat transfer from the combustion chamber wall was calculated depending on the engine crankshaft angle (Çelik, Örs et al. 2017).

$$\frac{dQ_w}{d\theta} = \left[S.h(T_g - T_w)\left(\frac{1}{6n}\right)\right] (J/°CA) \tag{6}$$

$$h_c = \left[\frac{130 \cdot P_c^{0.8}(\vartheta_p + 1{,}4)^{0.8}}{V^{0.06} T_g^{0.4}}\right] (W/m^2 K) \tag{7}$$

$$S = \left[\frac{V_2}{A_p}\pi . D_p + 2.A_p\right] (m^2) \tag{8}$$

S = Combustion chamber wall surface area (m^2)
h = Heat transfer coefficient (W/m^2K),
T_w = Combustion chamber wall surface temperature (K)
Pc = Cylinder pressure (bar)
ϑ_p = Average piston speed (m/s)

V_2 = Combustion chamber volume (m^3)

Error Analysis and Estimation of Uncertainty

The accuracy of the results and measured values obtained in experimental and applied studies is also very important. The most important factors affecting the accuracy of the results are the reading of the test values, the environmental values of the test environment and the errors caused by the test measuring equipment. Uncertainty analysis is a procedure that provides a scientific approach to the impact of these errors on results (Kanoglu 2000).

According to this method, the size to be measured in the system is R, and the n independent variables affecting this size are $x_1, x_2, x_3, \ldots, x_n$. In this case,

$$R = R(x_1, x_2, x_3, \ldots, x_n) \tag{9}$$

can be written. If the error rate of each independent variable is w1, w2, w3,...., and the error rate of wn and R is w_R, the equation according to the relation presented by Kline and McClintock is as follows: as:

$$W_{R,\max} = \left[\sum_{i=1}^{n} \left(W_{xi} \frac{\partial R}{\partial x_i} \right)^2 \right]^{1/2} \tag{10}$$

W_{xi} = The accuracy or the error value

Fixed error measurements are given in Table 5.

Table 5. Measurements and calculated parameters

No	Parameters	Unit	Uncertainty
1	Engine Power	%	5.21
2	Specific fuel consumption	%	6.68
3	Engine Speed	%	1.5
4	Length and diameter	mm	± 0.1
5	Stopwatch	second	± 0.5
6	Digital Scales	gram	± 0.01

Results and Discussion

Engine Power

Figure 2 shows the effect of standard diesel, biodiesel, and fuel mixtures on engine power at full load between 1000 – 4000 rpm engine speed. The maximum engine power was obtained as 44.44 kW at 3000 rpm for B7E5 fuel mixture whereas D100 as 40 kW at 3000 rpm. Besides, engine power was increased by B5E5 and B7E5 fuel mixtures averagely as 2.4% and 4.3% respectively, whereas B100 fuel decreased 3.2%, compared with the D100. According to the results obtained from the tests, B5E5 and B7E5 fuel mixtures at all engine speeds showed an increase in engine power compared to D100 fuel. The reason of this increase in engine power in B5E5 and B7E5 fuel mixtures with the addition of bioethanol into diesel fuel may have been due to improved fuel injection properties and atomization. That may be decreased in the density and viscosity of the mixtures. Moreover, bioethanol has higher oxygen content and high Hydrogen/Carbon (H/C) ratio which can positively affect the combustion process and cause the fuel to burn completely and the amount of oxygen and BSFC in the mixture to increase compared to D100. Moreover, bioethanol has higher H/C ratio and oxygen content. This may be positive effect on improving the combustion process and air resulting in complete combustion of fuel and the increase in the amount of oxygen and BSFC in the mixture, compared to D100. (Zhu, Cheung et al. 2011, Jamrozik, Tutak et al. 2017). Consequently, engine power for blends was increased. B100 fuel has low heat value, which lower the engine power.

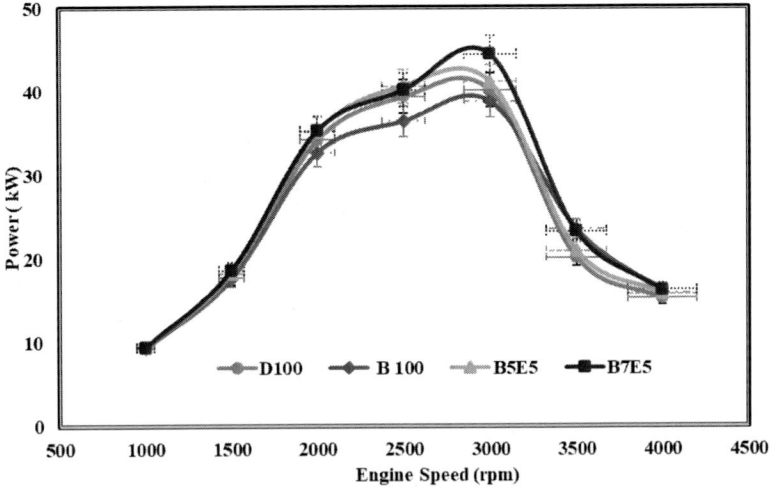

Figure 2. Engine power for diesel, biodiesel, and bioethanol fuel mixtures with respect to varying engine speed.

Engine Torque

Figure 3 are illustrated the engine torque graphic for diesel, biodiesel, and fuel mixtures between 1000 – 4000 rpm engine speed under full load. The maximum torque was obtained as 168 Nm for B7E5 fuel mixture while B5E5 fuel mixture as 166 Nm at 2000 rpm. Besides, engine torque was increased by B5E5 and B7E5 fuel mixtures averagely as 1% and 1.3% respectively, whereas B100 fuel decreased 9%, compared to D100.

As a result, this increase in engine torque in B5E5 and B7E5 fuel mixtures, bioethanol may have been helped to increase in value of peak pressure during combustion due to its very high latent heat of vaporization. Besides, the effect of the oxygen in the fuel can be explained by the acceleration of the combustion process and the increase of the temperature and pressure.

In addition, the high evaporation heat of bioethanol can be caused increases the volumetric efficiency of the engine during suction and compression time due to reduces in diffusion combustion and total combustion duration. As a result of all this, engine torque may have increased slightly. The lower heat value (LHV) of B100 fuel is lower than the value of diesel fuel, sothe engine torque value is decreased (Murcak, Haşimoğlu et al. 2015, Jamrozik, Tutak et al. 2017).

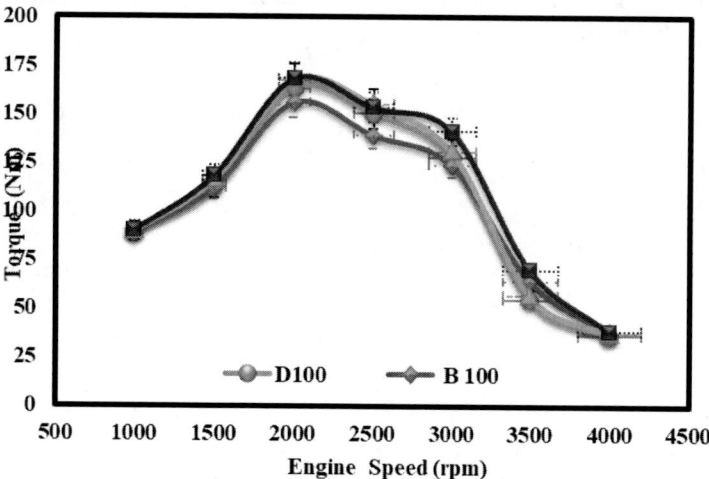

Figure 3. Engine torque for diesel, biodiesel, and bioethanol fuel mixtures with respect to varying engine speed.

Brake Thermal Efficiency (BTE)

BTE graphic for diesel, biodiesel and bioethanol fuel mixtures versus diesel fuel is given in Figure 4. The maximum BTE was obtained as 36.4% with B7E5 fuel mixture at 3000 rpm. B7E5 fuel mixtures increased averagely as 1% whereas B100 fuel decreased 4.8%, compared with the D100. The highest brake thermal efficiency for B100 fuel was obtained 33.6% at 3000 rpm.

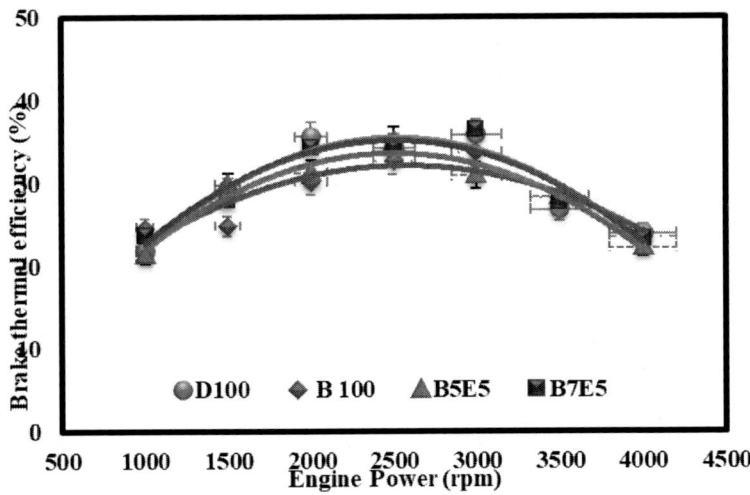

Figure 4. BTE for diesel, biodiesel, and bioethanol fuel mixtures with respect to varying engine speed.

Consequently, the reason of the lower BTE of B100 fuel is due to its higher viscosity, LHV and low evaporation of biodiesel. The high viscosity weakens the fuel atomization during the spraying process, which causes deposits and wear on fuel pump components and injectors as requires more energy to pump the fuel. Therefore, the efficiency may be reduced for B100 fuel (Kannan and Anand 2011). BTE for B7E5 fuel mixture increased. This may be due to the fact that the oxygen in the mixture improves combustion and the longer ignition delay which causes more fuel to accumulated and burned in the premixed mode (Subbaiah, Gopal et al. 2010). Another reason is that the addition of bioethanol causes better combustion and fuel improves atomization due to the reduction in the viscosity and density of the fuel. There was slightly decreased in B5E5 fuel mixture compared to B7E5 fuel mixture.

The reason of this may be from the B5E5 fuel mixture has a higher density than B7E5 fuel.

Brake Specific Fuel Consumption (BSFC-be)

BSFC is affected by the engine's volumetric fuel injection system, the density of the fuel and the heat value. In general, it is mostly affected by the lower heat value (Prbakaran and Viswanathan 2016). The results of the effect of BSFC diesel, biodiesel, and ethanol fuel mixtures between 1000 – 4000 rpm engine speed under full load are presented in Figure 5. The lowest BSFC was obtained as 219.42 g/kWh for B7E5, whereas B100 and B5E5 fuel blend obtained as 267.83g/kWh and 257.92 g/kWh respectively, comparing with 218.18 g/kWh the D100 fuel at 3000 rpm. Also, BSFC value for B100, B5E5 and B7E5 fuel mixtures were increased averagely 22.6%, 10% and 8% compared with D100 fuel respectively. The highest BSFC at all engine speeds was carried out with B100.

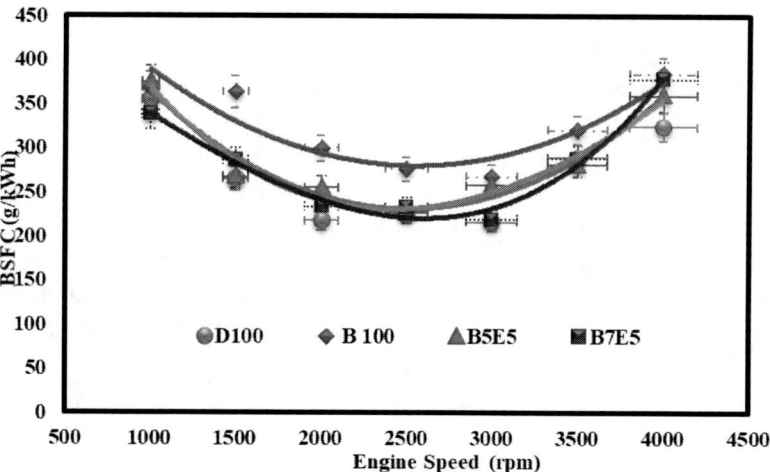

Figure 5. Bsfc for diesel, biodiesel, and bioethanol fuel mixtures with respect to varying engine speed.

Consequently, the reason for the high BSFC values of B100 fuel and blends is the low lower heating value of these fuels compared to D100 fuel. Another reason is the fuel increase of B100, B5E5 and B7E5 fuel and mixtures to achieve the same engine power compared to D100 fuel. Many authors reported that BSFC value of a fuel mixture increases due to the decrease of the low heat value of the mixture and these increases depend on the ratio of bioethanol and biodiesel content of the blends.

Another reason for the increase in BSFC may be due to changes the combustion timing and injection timing by biodiesel high cetane number. The reason for the decrease in BSFC for the other blend compared to B100 fuel was that the increase in combustion pressure due to the high lower heating value and high combustion efficiency and a slightly increase in fuel consumption caused a decrease in BSFC for the corresponding engine power.

Cylinder Pressure

The cylinder pressure changes according to the crank angle for diesel, biodiesel and bioethanol fuel mixtures versus diesel fuel is given in Figure 6. The maximum cylinder pressure was obtained as 104.25 bars with B7E5 fuel blends at 3000 rpm. In the same period, the peak pressures of 100.5, 100 and 103.2 bars were recorded for D100, B100 and B5E5 fuel blends, respectively. At 2000 rpm the maximum cylinder pressures 91.9, 92.10, 94.7 and 95.70 bar was obtained for D100, B100, B5E5 and B7E5 fuel blends, respectively.

This led to in longer combustion during the pre-mixed combustion phase and increased fuel accumulation due to longer ignition delay period. This may be increased heat release rate and cylinder pressure (Buyukkaya 2010, Hulwan and Joshi 2011, Çelik, Örs et al. 2017). As a result of this, the cylinder pressure may be slightly increased for B5E5 and B7E5 fuel blends comparison with D100 fuel and B100 fuel. Therefore, with the use of B100 fuel, the ignition delay period is slightly shortened due to injection and the ignition time starts early and causes the combustion phase to begin early. As a result, cylinder pressure and heat release rate start to rise earlier. But lower pressures arise during the expansion process due to the short burning time of biodiesel fuel. Therefore, the cylinder pressure for the B100 may have slightly decreased compared to the D100 fuel. For another reason, the cylinder peak pressure for B100 may be lower owing to low volatility, high viscosity, poor atomization, poor mixture formation and coarse spray formation. Therefore, peak cylinder pressure for D100 fuel may have increased fuel due to long ignition delay causes more fuel to burned comparison with B100 fuel.

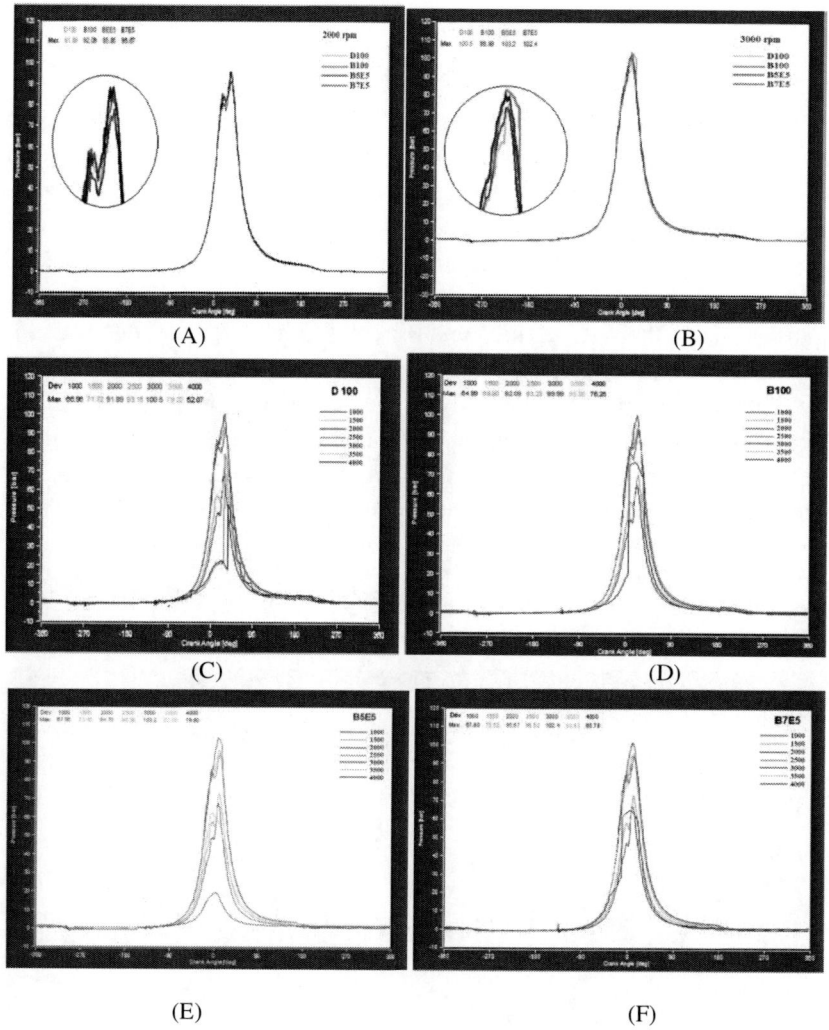

Figure 6 (A-F). Pressure traces of engine powered by diesel-biodiesel-bioethanol fuel mixtures.

Heat Release Rate (HRR)

The HRR changes according to the crank angle for diesel, biodiesel and bioethanol fuel mixtures versus diesel fuel is given in Figure 7. It is observed that the highest heat release rates were obtained as 524.94 kj/m^{3}deg and 598.3

kj/m³deg with B7E5 fuel blends at 2000 and 3000 rpm, respectively. HRR of 576.12, 492.06 and 515.4 kj/m³deg were recorded for D100, B100 and B5E5 fuel blends, respectively.

According to the HRR results, it has been shown that the HRR increased with the addition of bioethanol to fuel mixtures. The use of fuel containing oxygen increases the combustion rate for all fuel mixtures and has a faster laminar flame rate than diesel. This may be the reason for the increase in HRR (V. Gnanamoorthia and Devaradjane 2013). Also, bioethanol in B5E5 and B7E5 fuel blends can reduce the viscosity of and increases latent heat of vaporization of the fuel blend which increases heat release rate and led to lengthen the ignition delay period. Therefore, the amount of fuel burned in the premixed mode increases and rapid combustion occurs and improved diffusive combustion at full loads. Consequently, the heat release rate of B5E5 and B7E5 fuel mixtures was higher than D100 fuel in all cycles. B100 has more oxygen content than diesel which the higher fuel–air ratio. But the high density and viscosity of biodiesel affects fuel atomization and slows down the air- fuel mixture ratio and uneven burning, hence lower HRR (Abedin, Masjuki et al. 2014).

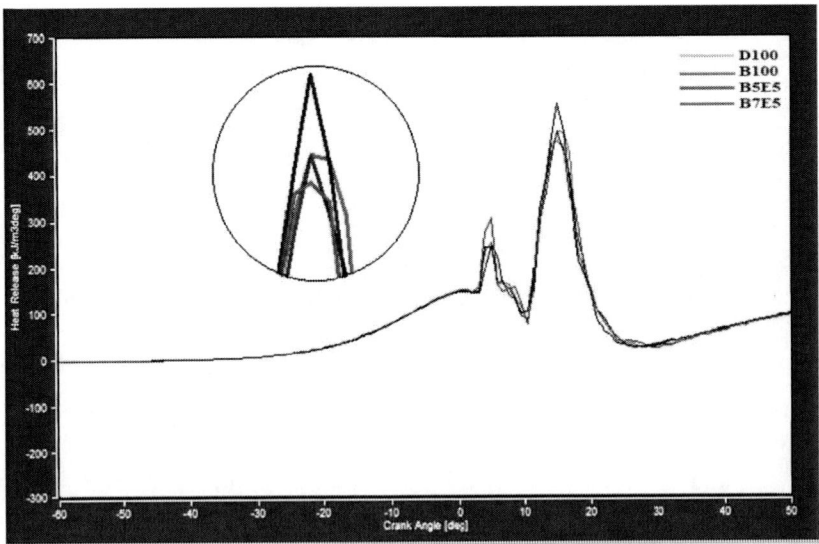

Figure 7. HRR traces of engine powered by diesel-biodiesel –bioethanol fuel mixtures.

Moreover, the high cetane number of biodiesels facilitates auto-ignition and shortens the premixed combustion phase which combustion start early. Therefore, B100 fuel decreases amount of fuel in the premixed phase combustion phase compared to the D100 fuel. Shorter premixed period provides lower peak cylinder pressure and lower HRR. B5E5 fuel blend exhibits slightly lower HRR than the B7E5 fuel blend. B7E5 fuel blend has a little higher oxygen content and calorific value than B5E5 which raises heat release rate.

Carbon Monoxide Emission

Figure 8 presents the carbon monoxide values for diesel, biodiesel, and bioethanol fuel mixtures between 1000 – 4000 rpm engine speed under full load. The CO emissions for the B100, B5E5 and B7E5 fuel blends decreased. Moreover, the minimum CO emission value was obtained as 0.01% with B100 fuel at all engine speed. At 2000 and 3000 rpm the CO emission values 0.008%, 0.064%, 0.023% and 0.009%, 0.015%, 0.015% was obtained for B100, B5E5 and B7E5 fuel blends compared with the D100 as 0.543% and 0.106%, respectively. Also, CO emission value for B100, B5E5 and B7E5 fuel mixtures were decreased averagely 82.8%, 42.97% and 52.7% compared with D100 fuel at all engine speed, respectively.

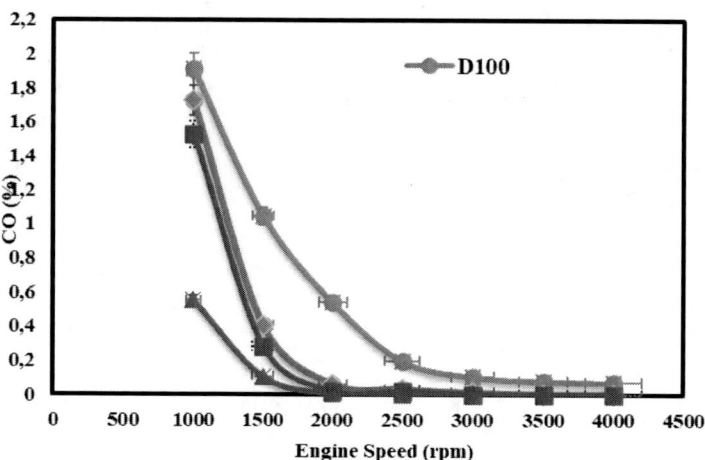

Figure 8. Carbon monoxide values for diesel, biodiesel, and bioethanol fuel mixtures with respect to varying engine speed.

As a result, it is observed that CO emissions in B100, B5E5 and B7E5 fuel mixtures decreased for all values of engine speed compared to D100 fuel. Also, CO emissions remarkably reduced for B5E5 and B7E5 fuel blends that would be due the presence of large amount of the oxygen content of bioethanol and biodiesel and improved combustion process owing to better air-fuel mixing which leads to more complete combustion. Another reason for low CO emissions could be the high cetane number shortening the ignition delay and allows a longer combustion duration (Abedin, Masjuki et al. 2014). Moreover, CO emissions are reduced as the turbocharger sends more air when it starts operating at high engine speeds and provides better combustion in all fuel mixtures. Similar results achieved in Subbaiah, Gopal, Hussain, Prasad and Reddy [39] studies.

Carbon Dioxide Emission

Carbon dioxide emission is a normal product the of hydro-carbon fuel combustion. When the hydro-carbon fuel is burned, it produces only CO_2 and water (H_2O) (Hulwan and Joshi 2011). Figure 9 are presented the carbon dioxide values for diesel, biodiesel, and bioethanol fuel mixtures between 1000 – 4000 rpm engine speed under full load. For the B100, B5E5 and B7E5 fuel blends, the CO_2 emissions increase compared to D100 fuel. At 2000 and 3000 rpm the CO_2 emission values 9.37%, 9.93%, 9.64% and 8.33%, 9.25%, 9.1% was obtained for B100, B5E5 and B7E5 fuel blends compared with the D100 as 9.04% and 8.82%, respectively. Also, CO_2 emission value for B100, B5E5 and B7E5 fuel mixtures were increased averagely 2.1%, 1.8% and 3.04% compared with D100 fuel at all engine speed, respectively.

The BSFC of B5E5 and B7E5 fuel mixtures is higher than D100 fuel. This may have increased the oxygen/fuel ratio and therefore slightly increased CO_2 emissions. Moreover, bioethanol contains less carbon and more oxygen than diesel which increases the oxygen/fuel ratio. This may improve the combustion process, which slightly increases CO_2 emissions. Enough oxygen for C/H atoms may be provided in the B100 fuel and thus the CO_2 emission values may be decreased for carbon atoms release from the exhaust as forms of CO_2. Therefore, the CO_2 emissions of B100 fuel may be slightly higher than D100 fuel on average. This result is an indication that the oxygen contained in biodiesel has a positive effect on combustion (Alptekin, Canakci et al. 2015). Similar results were obtained with the study prepared by Shi, Pang et al. (2006).

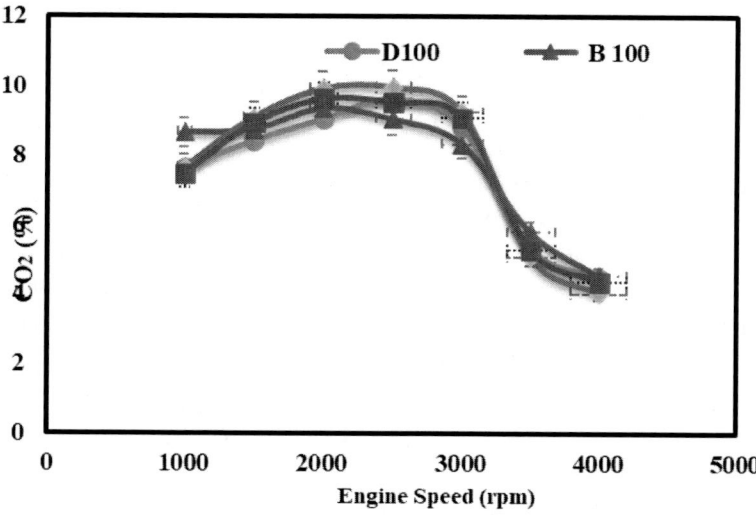

Figure 9. Carbon dioxide values for diesel, biodiesel, and bioethanol fuel mixtures with respect to varying engine speed.

Hydrocarbon Emission (HC)

HC emissions are indicators of efficiency of combustion or completeness and unburned hydrocarbon in the exhaust emissions. HC concentrations in the engine exhaust are reduced when it is closer to stoichiometric burning or complete combustion (Zhu, Cheung et al. 2010). Figure 10 are presented the HC emission values for diesel, biodiesel and bioethanol fuel mixtures between 1000 – 4000 rpm engine speed under full load. For the B100, B5E5 and B7E5 fuel blends, HC emissions decrease at all engine speed compared with D100 fuel. Minimum HC emissions were obtained as 7 ppm with B100 fuel while D100 fuel as 10 ppm at 1500 rpm. At 2000 and 3000 rpm the HC emission values 7 ppm, 7 ppm, 9 ppm and 7 ppm, 7 ppm, 9 ppm was obtained for B100, B5E5 and B7E5 fuel blends compared with the D100 as 12 ppm and 14 ppm, respectively. Also, HC emission value for B100, B5E5 and B7E5 fuel mixtures were increased averagely 51%, 33%, and 26% compared with D100 fuel at all engine speed, respectively. Maximum HC emissions were obtained as 22 ppm with D100 fuel. HC emissions may be reduced as the high cetane number (CN) of biodiesel due to it shortens the ignition period and improvers combustion. Also, B5E5, B7E5 and B100 fuel and blends have more oxygen molecules than D100 fuel. This can contribute positively to the development of combustion and lead to reduces emissions. B5E5 and B7E5 blends showed slightly higher HC emissions due to longer in ignition delay period all engine speeds. Another reason is that decreases the viscosity and density of the mixture by adding a small amount of bioethanol. Fuel blends can improve spray properties and atomization due to lower viscosity and density, which can improve combustion by making a slight cooling effect in the combustion chamber (Zhu, Cheung et al. 2010). This can lead to a reduction in HC emissions. Similar results obtained Barabás, Todoruţ et al. (2010) which stated that HC emission reduced by the addition of 5% bioethanol to fuel blends.

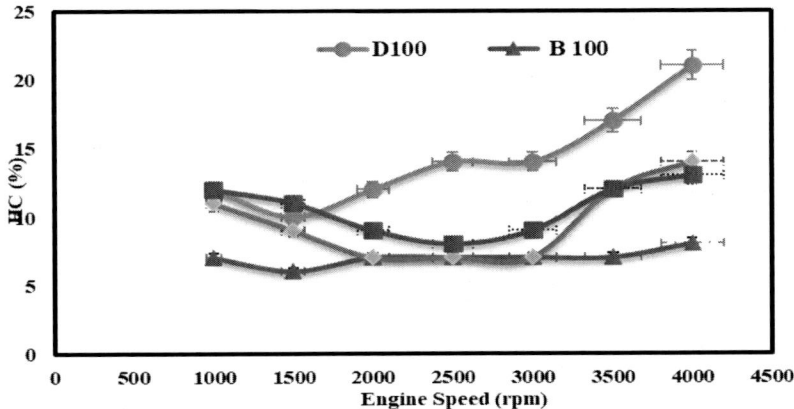

Figure 10. HC values for diesel, biodiesel, and bioethanol fuel mixtures with respect to varying engine speed.

Nitrogen Oxides Emission (NO_x)

NO_x formation is a major function of temperature during the combustion and composes through the reaction of N_2 and O_2 when the combustion chamber temperature rises above 1800 °K. This high temperature is caused by better mixing of large fuel quantity and air. In addition, some parameters such as viscosity, cetane number, the amount of oxygen in fuel and reaction time also affect the NOx emissions Alptekin, Canakci et al. (2015). Figure 11 are presented the NO_x emission values for diesel, biodiesel, and bioethanol fuel mixtures between 1000 – 4000 rpm engine speed under full load. For the B100, B5E5 and B7E5 fuel blends, NO_x emissions increase at all engine speed compared with D100 fuel. Minimum NO_x emissions were obtained as 412 ppm with D100 fuel. At 2000 and 3000 rpm the HC emission 1120 ppm, 951 ppm, 952 ppm and 1342 ppm, 1295 ppm, 1301 ppm was obtained for B100, B5E5 and B7E5 fuel blends compared with the D100 fuel as 963 ppm with 1250 ppm respectively. Also, NO_x emission value for B100, B5E5 and B7E5 fuel mixtures were increased averagely 14.29%, 1.4% and 2.94% compared with D100 fuel at all engine speed, respectively. Maximum NO_x emissions were obtained as 1342 ppm with B100 fuel.

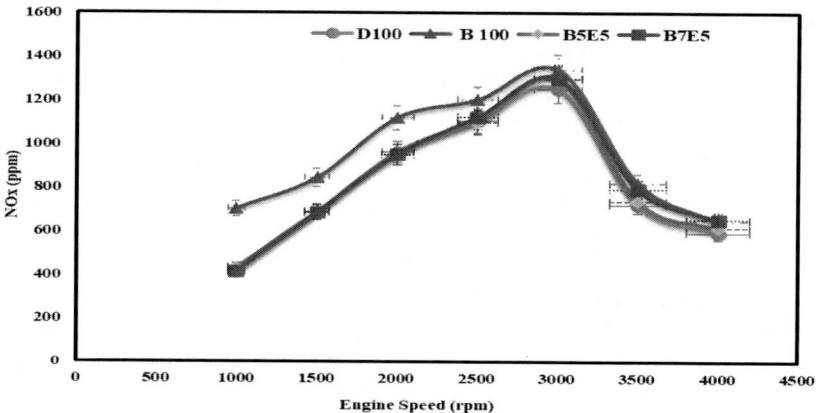

Figure 11. NO_x values for biodiesel bioethanol and diesel, fuel mixtures with respect to varying engine speed.

The reason for the slightly increase of NOx emissions in B7E5 fuel is due to the higher HRR and combustion temperature. The high temperature and pressure may create larger regions of close to stoichiometric burned gas, NO_x emissions may be increased. The lowering of the cetane number of bioethanol mixtures increases the ignition delay and causes more fuel to accumulate in the premixed combustion period. Therefore, there is more energy to vaporize alcohols, which improves the combustion, resulting in a rapidly increased the rate of heat release, lead to the higher NO_x emissions. (Hulwan and Joshi 2011). Also, the latent evaporation feature of bioethanol lowers the combustion temperature by cooling effect and causes slightly lower NOx emissions compared to B100 fuel. Obviously, the oxygen content in the biodiesel caused an increase in the temperature of the combustion. That causes increased NOx emissions (Köse and Acaroğlu 2020). Similar results obtained by Barabás, Todoruţ et al. (2010).

Conclusion

This paper was investigated the effect of on performance, combustion, and exhaust emission characteristics of using biodiesel, bioethanol, and diesel fuels in a diesel engine. The following results were obtained.

For engine power value of B5E5 and B7E5 fuel mixtures increased averagely as 2.4% and 4.3% whereas B100 fuel decreased 3.2%, compared with the D100 fuel at all engine speed, respectively. The engine power of the blended fuels increases with additional of bioethanol in the fuel due to the higher oxygen concentrations. For the mixtures of B7E5 fuel, maximum engine power was obtained as 44.44 kW. Engine torque increased averagely as 1% and 1.3% for B5E5 and B7E5 mixtures in whereas B100 fuel decreased 9%, comparison to diesel fuel. Viscosity, density, amount of oxygen and cetane number of the fuel mixture played vital roles in engine performance.

The highest BTE value was obtained as 36.4% with B7E5 fuel mixture at 3000 rpm. B7E5 fuel mixtures increased averagely as 1% whereas B100 fuel decreased 4.8%, compared with the D100. BSFC increased averagely as 22.65% 9.9% and 8% for B100, B5E5 and B7E5 mixtures compared with the D100 fuel. The peak pressures of 100.5, 100 and 103.2 bars were recorded for D100, B100 and B5E5 fuel blends, respectively. Maximum the cylinder pressure value was obtained as 104.25 bar with B7E5 fuel blends at 3000 rpm.

The maximum heat release rates were obtained as 598.3 kj/m^{3}deg with B7E5 fuel blends at 3000 rpm. CO emission value for B100, B5E5 and B7E5 fuel blends were decreased averagely 82.8%, 42.97% and 52.7% compared with D100 fuel. Bioethanol fuel blends provided higher CO_2 emissions than diesel fuel. CO_2 emission value for B100, B5E5 and B7E5 fuel mixtures were increased averagely 2.1%, 1.8% and 3.04% compared with D100 fuel. For the B100, B5E5 and B7E5 fuel blends, HC emissions decrease at all engine speed compared with D100 fuel. Minimum HC emissions were obtained as 7 ppm with B100 fuel. HC emission value for B100, B5E5 and B7E5 fuel mixtures were increased averagely 51%, 33%, and 26% compared with D100 fuel at all engine speed.

NO_x emission value for B100, B5E5 and B7E5 fuel mixtures were increased averagely 14.29%, 1.4% and 2.94% compared with D100 fuel. Maximum NO_x emissions were obtained as 1342 ppm with B100 fuel.

Based on all the results of the present experimental study, bioethanol-biodiesel blends can replace the CI engine applications out of any modifications on the engine thereby reducing the dependency of fossil fuels. It can be concluded that the *Cynara cardunculus seed oil* can be an effective and inexpensive oxygenated fuel due to the many important benefits mentioned above.

Overall, it can be concluded that ethanol-biodiesel blends are more effective for emission reduction, except for nitrogen oxide emissions. The engine test results showed that engine performance is not adversely affected by *cynara cardunculus seed oil biodiesel*. Besides, It can be used in non-edible vegetable-based biodiesel and bioethanol fuels as a fuel blend with diesel fuel in diesel engines. According to this result, introducing biodiesel and bioethanol fuel can help increase biofuel consumption from the diesel segment.

Acknowledgments

This study was supported by Scientific Research Project Coordinators of Selcuk University under the project number of 16101008.

References

Abedin, M. J., H. H. Masjuki, M. A. Kalam, A. Sanjid, S. M. A. Rahman and I. M. R. Fattah (2014). "Performance, emissions, and heat losses of palm and jatropha biodiesel blends in a diesel engine." *Industrial Crops and Products* 59: 96-104.

Alptekin, E., M. Canakci, A. N. Ozsezen, A. Turkcan and H. Sanli (2015). "Using waste animal fat based biodiesels–bioethanol–diesel fuel blends in a DI diesel engine." *Fuel* 157: 245-254.

An, H., W. M. Yang and J. Li (2015). "Effects of ethanol addition on biodiesel combustion: A modeling study." *Applied Energy* 143: 176-188.

Armas, O., S. Martínez-Martínez and C. Mata (2011). "Effect of an ethanol–biodiesel–diesel blend on a common rail injection system." *Fuel Processing Technology* 92(11): 2145-2153.

Aydin, H. and C. İlkılıç (2010). "Effect of ethanol blending with biodiesel on engine performance and exhaust emissions in a CI engine." *Applied Thermal Engineering* 30(10): 1199-1204.

Barabás, I., A. Todoruţ and D. Băldean (2010). "Performance and emission characteristics of an CI engine fueled with diesel–biodiesel–bioethanol blends." *Fuel* 89(12): 3827-3832.

Buyukkaya, E. (2010). "Effects of biodiesel on a DI diesel engine performance, emission and combustion characteristics." *Fuel* 89(10): 3099-3105.

Çelik, M., İ. Örs, C. Bayindirli and M. Demiralp (2017). "Experimental investigation of impact of addition of bioethanol in different biodiesels, on performance, combustion

and emission characteristics." *Journal of Mechanical Science and Technology* 31(11): 5581-5592.
Curt, M. D., G. Sanchez and J. Fernandez (2002). "The potential of Cynara cardunculus L. for seed oil production in a perennial cultivation system." *Biomass and Bioenergy* 23: 33-46.
de Oliveira, A., O. S. Valente and J. R. Sodré (2018). "Effects of ethanol addition to biodiesel-diesel oil blends (B7 and B20) on engine emissions and fuel consumption." *MRS Advances* 2(64): 4005-4015.
Fathi, M., R. Khoshbakhti Saray and M. David Checkel (2010). "Detailed approach for apparent heat release analysis in HCCI engines." *Fuel* 89(9): 2323-2330.
Gnanamoorthia, V. and G. Devaradjane (2013). "Effect of diesel-ethanol blends on performance, combustion and exhaust emission of a diesel engine." *International Journal of Current Engineering and Technology* 3(1): 36-42.
Hulwan, D. B. and S. V. Joshi (2011). "Performance, emission and combustion characteristic of a multicylinder DI diesel engine running on diesel–ethanol–biodiesel blends of high ethanol content." *Applied Energy* 88(12): 5042-5055.
Jamrozik, A., W. Tutak, M. Pyrc and M. Sobiepański (2017). "Effect of diesel-biodiesel-ethanol blend on combustion, performance, and emissions characteristics on a direct injection diesel engine." *Thermal Science* 21(1 Part B): 591-604.
Kannan, G. R. and R. Anand (2011). "Combustion characteristics of a diesel engine operating on biodiesel–diesel–ethanol mixtures." *Proceedings of the Institution of Mechanical Engineers, Part A: Journal of Power and Energy* 225(8): 1076-1087.
Kanoglu, M. (2000). "Uncertainty analysis of cryogenic turbine efficiency." *Mathematical and Computational Applications* 5: 169–177.
Köse, H. and M. Acaroğlu (2020). "The effect of hydrogen addition to cynara biodiesel on engine performance and emissions in diesel engines." *Energy Sources, Part A: Recovery, Utilization, and Environmental Effects*.
Köse, H. and M. Ciniviz (2013). "An experimental investigation of effect on diesel engine performance and exhaust emissions of addition at dual fuel mode of hydrogen." *Fuel Processing Technology* 114: 26-34.
Kwanchareon, P., A. Luengnaruemitchai and S. Jai-In (2007). "Solubility of a diesel–biodiesel–ethanol blend, its fuel properties, and its emission characteristics from diesel engine." *Fuel* 86(7-8): 1053-1061.
Lee, W.-J., Y.-C. Liu, F. K. Mwangi, W.-H. Chen, S.-L. Lin, Y. Fukushima, C.-N. Liao and L.-C. Wang (2011). "Assessment of energy performance and air pollutant emissions in a diesel engine generator fueled with water-containing ethanol–biodiesel–diesel blend of fuels." *Energy* 36(9): 5591-5599.
Micic, V. and M. Jotanovic (2015). "Bioethanol as fuel for internal combustion engines." *Zastita materijala* 56(4): 403-408.
Murcak, A., C. Haşimoğlu, İ. Çevik and H. Kahraman (2015). "Effect of injection timing to performance of a diesel engine fuelled with different diesel–ethanol mixtures." *Fuel* 153: 569-577.
Nantha Gopal, K., B. Ashok, K. Senthil Kumar, R. Thundil Karuppa Raj, S. Denis Ashok, V. Varatharajan and V. Anand (2017). "Performance analysis and emissions profile of cottonseed oil biodiesel–ethanol blends in a CI engine." *Biofuels* 9(6): 711-718.

Ozcelik, A. E., M. Acaroglu, H. Aydogan and M. Cinar (2018). Effect of common-rail diesel engine bioethanol-biodiesel-eurodiesel mixtures on engine performance and emissions. 4. *International Conference On Enviromental Science and Technology (ICOEST)*. Ukraine: 226-232.

Parthasarathy, M., J. Isaac JoshuaRamesh Lalvani, B. Dhinesh and K. Annamalai (2016). "Effect of hydrogen on ethanol-biodiesel blend on performance and emission characteristics of a direct injection diesel engine." *Ecotoxicol Environ Saf* 134(Pt 2): 433-439.

Pasqualino, J. C. (2006). *Cynara Cardunculus as an Alternative Crop for Biodiesel Production* [Doctoral, Universitat RoviraI Virgili.]

Pidol, L., B. Lecointe, L. Starck and N. Jeuland (2012). "Ethanol–biodiesel–Diesel fuel blends: Performances and emissions in conventional Diesel and advanced Low Temperature Combustions." *Fuel* 93: 329-338.

Piscioneri, I., N. Sharma, G. Bavielle and S. Orlandini (2000). "Promising industrial energy crop, Cynara cardunculus a potential source for biomass production and alternative energy." *Energy Conversion & Management* 41: 1091-1105.

Prbakaran, B. and D. Viswanathan (2016). "Experimental investigation of effects of addition of ethanol to bio-diesel on performance, combustion and emission characteristics in CI engine." *Alexandria Engineering Journal*.

Rakopoulos, D. C. (2012). "Heat release analysis of combustion in heavy-duty turbocharged diesel engine operating on blends of diesel fuel with cottonseed or sunflower oils and their bio-diesel." *Fuel* 96: 524-534.

Shi, X., X. Pang, Y. Mu, H. He, S. Shuai, J. Wang, H. Chen and R. Li (2006). "Emission reduction potential of using ethanol–biodiesel–diesel fuel blend on a heavy-duty diesel engine." *Atmospheric Environment* 40(14): 2567-2574.

Subbaiah, G. V., K. R. Gopal and S. A. Hussain (2010). "The effect of biodiesel and bioethanol blended diesel fuel on the performance and emission characteristics of a direct injection diesel engine." *Iranica Journal of Energy & Environment* 1 3: 211-221.

Sukjit, E., J. M. Herreros, K. D. Dearn, A. Tsolakis and K. Theinnoi (2013). "Effect of hydrogen on butanol–biodiesel blends in compression ignition engines." *International Journal of Hydrogen Energy* 38(3): 1624-1635.

Teoh, Y. H., K. H. Yu, H. G. How and H. T. Nguyen (2019). "Experimental investigation of performance, emission and combustion characteristics of a common-rail diesel engine fuelled with bioethanol as a fuel additive in coconut oil biodiesel blends." *Energies* 12(10).

Tse, H., C. W. Leung and C. S. Cheung (2015). "Investigation on the combustion characteristics and particulate emissions from a diesel engine fueled with diesel-biodiesel-ethanol blends." *Energy* 83: 343-350.

Yilmaz, N. (2012). "Comparative analysis of biodiesel–ethanol–diesel and biodiesel–methanol–diesel blends in a diesel engine." *Energy* 40(1): 210-213.

Yilmaz, N., F. M. Vigil, A. Burl Donaldson and T. Darabseh (2014). "Investigation of CI engine emissions in biodiesel–ethanol–diesel blends as a function of ethanol concentration." *Fuel* 115: 790-793.

Zhu, L., C. S. Cheung, W. G. Zhang and Z. Huang (2010). "Emissions characteristics of a diesel engine operating on biodiesel and biodiesel blended with ethanol and methanol." *Sci Total Environ* 408(4): 914-921.

Zhu, L., C. S. Cheung, W. G. Zhang and Z. Huang (2011). "Combustion, performance and emission characteristics of a DI diesel engine fueled with ethanol–biodiesel blends." *Fuel* 90(5): 1743-1750.

Chapter 4

The Study of Properties in Biodiesel/Butanol and Biodiesel/Diesel/Butanol Blends

S. D. Romano[*], PhD
Renewable Energy Group (GER),
Department of Mechanical Engineering,
Faculty of Engineering,
University of Buenos Aires, and CONICET, Argentina
Ciudad Autónoma de Buenos Aires, Argentina

Abstract

Biodiesel/butanol and biodiesel/diesel/butanol blends have recently received international attention due to the improvement in their properties because of the addition of butanol to the fuel/s, and the possibility of massive use of such systems in a near future. The data about these blends in the scientific literature are mainly related to the study of fuel consumption, combustion performance, and exhaust emissions (NOx, CO, HC, and smoke) in diesel engines. However, most studies also include the determination of some properties of technological interest in a few samples.

This chapter presents the study of two important properties - flash point (related to safety during transportation and storage of the blends) and refractive index (an optical property of translucent substances that can give an account of their quality) - in biodiesel/butanol and biodiesel/diesel/butanol blends. The flash point has been very poorly studied in these systems, whereas the refractive index has not been studied.

---

[*] Corresponding Author's Email: sromano@fi.uba.ar.

In: The Future of Biodiesel
Editor: Michael F. Simpson
ISBN: 979-8-88697-166-8
© 2022 Nova Science Publishers, Inc.

Experimental results show that, at low butanol concentrations in the blends, the flash point values decrease sharply, whereas above 20% butanol content in volume in the blends, they remain practically constant. Refractive index values decrease linearly with increasing butanol content in the blends and temperature.

Keywords: biodiesel, butanol, diesel, blends, refractive index, flash point

Introduction

Most of the energy worldwide comes from fossil fuels, which are limited, non-renewable and polluting resources. Additionally, the energy demand increases as the world population grows. Fossil fuels are used in numerous activities, including the industries, agricultural machinery, vehicles, and different kinds of transportation. However, two decades ago, pioneering countries started blending diesel (produced from petroleum) with a liquid biofuel (produced from biomass) for automotive use: biodiesel (Knothe et al., 2005; Romano et al., 2005; Knothe, 2005). Biodiesel thus emerges as a renewable, less polluting alternative, and is of particular interest in highly petroleum-dependent countries that do not have enough oil reserves to be self-sufficient. The proportion of biodiesel in the mixture with diesel depends on the legislation of each country. However, the biodiesel concentration in the blend is usually up to 20% in volume (v/v).

On the other hand, since alcohols provide a higher amount of oxygen, diminishing the particulate emissions (Lapuerta et al., 2010), in the last years, researchers have been testing higher alcohols, such as butanol, as new alternative renewable liquid fuels to improve the properties of their blends with biodiesel and diesel/biodiesel.

Analytical Framework

Literature Review

Several researchers, such as Campos-Fernández et al. (2012), Siwale et al. (2013), Yoshimoto et al. (2013), Tüccar et al. (2014), Wei et al. (2014), Yilmaz and Vigil (2014), Atmanli (2016a and 2016b), Atmanli and Yilmaz (2018), among others, have carried out studies on the addition of alcohols with different carbon chain length to biodiesel, diesel and biodiesel/diesel blends, and tested their response in diesel engines.

Biobutanol, one of the most important higher alcohols that can be classified as second-generation biofuel, can be used as a partial substitute for biodiesel or diesel (Nanthagopal et al., 2018) and is thus useful for the mitigation of environmental pollution.

Studies performed by Lapuerta et al. (2010 and 2018), Babu et al. (2017), and Yilmaz and Atmanli (2017) have shown that, at temperatures higher than 0°C, diesel/butanol and biodiesel/butanol blends remain stable for several days without phase separation, in any proportion. Several researchers have also shown that butanol can replace up to 50% of diesel, improving its performance and emissions when used as fuel in diesel engines (Babu et al., 2017; Altun et al., 2011; Lebedevas et al., 2010; Valentino et al., 2012). In addition, Babu et al. (2017), Altun et al. (2011), and Yilmaz and Vigil (2014) have shown that butanol can be mixed up to 20% with biodiesel, replacing diesel to be used in diesel engines without any modification. Compared to biodiesel and diesel, butanol/biodiesel blends produce lower emissions of elemental carbon, particulate matter, volatile hydrocarbons, and polycyclic aromatics. According to Imdadul et al. (2016a), the addition of butanol to biodiesel reduces the viscosity of the blend improving its atomization and the efficiency of the combustion process. Babu et al. (2017) showed that the addition of butanol to the diesel/biodiesel blend, even at low concentrations, has substantial benefits for human health, vegetation, and animal life.

In studies of biodiesel/butanol and biodiesel/diesel/butanol blends, several researchers, including Tüccar et al. (2014), Imdadul et al. (2016a and 2016b), Işık et al. (2017), Atmanli (2016a and 2016b), Nanthagopal et al. (2018a), Yasin et al. (2017), Geacai et al. (2017), Kuszewski (2018), Çelebi and Aydın (2018) and Tipanluisa et al. (2021), measured some characterization properties required by international standards, such as cetane number, density, viscosity, caloric value, cold filter plugging point, cloud point, oxidation stability, and flash point. In particular, the flash point is related to safety during transportation and storage of the fuel mixture. However, it should be noted that the flash point of long-chain alcohol/diesel or biodiesel binary systems has been poorly studied. In addition to these properties, alternative properties such as the refractive index, which is an optical property that can be easily measured with high precision, can give an account of the fuel or blend. Although the refractive index is a property with many applications (Alviso and Romano, 2021), this property has not been studied in the mentioned systems. The technique to determine the refractive index has many advantages: it is non-destructive and precise, the equipment is relatively inexpensive and requires very small volume of samples, and the test conditions are safe (Alviso et al., 2020).

Given the excellent properties conferred by the addition of butanol to diesel and biodiesel, and the possibility of massive use of these promising systems in a near future, it is important to evaluate these blends through a systematic study of their properties. This chapter presents the results of the study of the flash point and refractive index as a function of temperature, in biodiesel/butanol and biodiesel/diesel/butanol blends.

Method

Samples

Soy biodiesel and diesel (D500) - both provided by a local refinery (YPF) - and commercial normal butanol - Anedra, 99.9% purity - were used to prepare the samples.

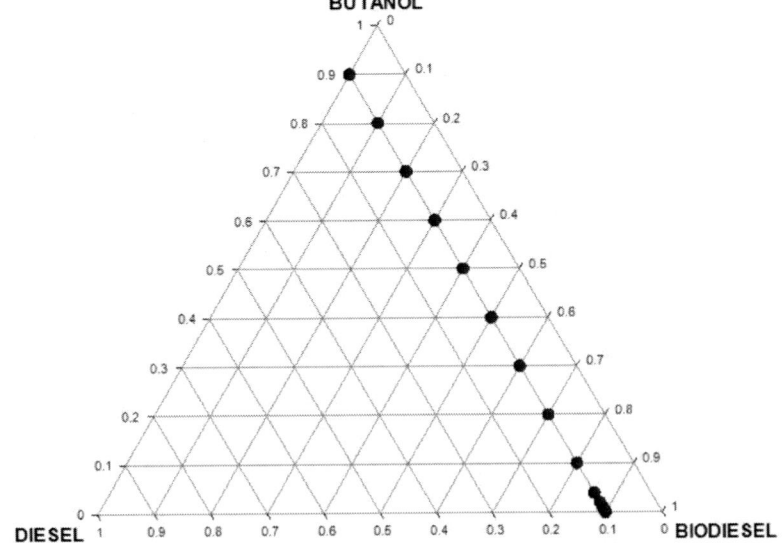

Figure 1. Concentrations of the ternary biodiesel/diesel/butanol system.

Biodiesel/butanol (BD/B) and biodiesel/diesel/butanol (BD/D/B) blends were prepared volumetrically (%v/v). For the binary system, the full composition range was studied, with greater emphasis on low concentrations of B. Indeed, the range between 0.5 and 5% of B in BD was measured in steps of 0.5% of B content, whereas the ranges between 5 and 20%, and between 20 and 100% of B in BD were measured in steps of 5 and 10% of B content, respectively. For the ternary system, D concentration was kept constant at 10% whereas the concentrations of BD and B were changed in steps of 10% between 0 and 90% of BD (90 and 0% of B, respectively). The concentration of B in the low range of the ternary blends (up to 10%) was studied deeply (0.2, 0.5, 1.0, 2.0, 4.0 and 7.0%). The concentration values of the ternary systems studied in this chapter can be seen in the ternary diagram shown in Figure 1.

Procedure and Equipment

The flash point (FP) of each system was measured according to procedure A of the ASTM D93 standard (2016), by using an electric Pensky Martens close cup equipment. The procedure basically consists in the constant heating (6°C/min) of a sample (70 ml) with continuous stirring. A small flame or incandescent object should be applied to the cup containing the sample at regular temperature intervals (1°C) with simultaneous interruption of the stirring. The FP corresponds to the lowest temperature at which the application of the flame generates the ignition of the vapor on the sample (ASTM D93, 2016).The refractive index (RI) was measured using an Abbe-type refractometer. Measurements were performed following the methodology indicated in the ASTM D1218 standard (2016). Monochromatic light of 589 nm was used, and the temperature was kept constant with a precision of +/- 0.1°C. The RI of each sample was studied at temperatures between 15 and 40°C with increments of 5°C.

Results and Discussion

Flash Point in Biodiesel/Butanol and Biodiesel/Diesel/Butanol Blends

Although the FP of BD/B (Yasin et al., 2017; Nanthagopal et. al, 2018a; Çelebi & Aydin, 2018), D/B (Campos-Fernández et al., 2012; Imbadul et al., 2016a;

Yang et al., 2017; Atmanli and Yilmaz, 2018; Kuszewski, 2018; Álvarez et al., 2019; Tipanluisa et al., 2021), and BD/D/B (Tüccar et al., 2014; Imbadul et al., 2016a; Imbadul et al., 2016b; Atmanli, 2016; Işık et. al, 2017; Çelebi and Aydin, 2018) blends has been previously studied, the number of samples studied by these researchers is low. The volumetric concentration of BD, D and B in these previously studied binary and ternary systems, as well as the authors, publication year, and BD origin (when it corresponds) are indicated in Table 1. It should also be noted that there are no systematic studies of the FP in BD/B or BD/D/B systems, and that the range of low alcohol concentrations (lower than 5%) has not been explored.

Table 1. Volumetric concentration of biodiesel, diesel, and butanol in the blends (literature data)

Authors	BD origin	BD (%v/v)	D (%v/v)	B (%v/v)	Year
Yasin et al.	Palm oil	90	-	10	2017
Nanthagopal et al.	Calophyllum Inophyllum oil	60	-	40	2018a
Nanthagopal et al.	Calophyllum Inophyllum oil	50	-	50	2018a
Nanthagopal et al.	Calophyllum Inophyllum oil	40	-	60	2018a
Çelebi and Aydin	Safflower oil	90		10	2018
Çelebi and Aydin	Safflower oil	80	-	20	2018
Campos-Fernández et al.	-	-	90	10	2012
Campos-Fernández et al.	-	-	85	15	2012
Campos-Fernández et al.	-	-	80	20	2012
Campos-Fernández et al.	-	-	75	25	2012
Campos-Fernández et al.	-	-	70	30	2012
Imbadul et al.	-	-	85	15	2016a
Imbadul et al.	-	-	80	20	2016a
Yang et al.	-	-	90	10	2017
Yang et al.	-	-	80	20	2017
Yang et al.	-	-	70	30	2017
Yang et al.	-	-	60	40	2017
Yang et al.	-	-	50	50	2017
Yang et al.	-	-	40	60	2017
Yang et al.	-	-	30	70	2017
Yang et al.	-	-	20	80	2017
Yang et al.	-	-	10	90	2017
Atmanli and Yilmaz	-	-	95	5	2018

Authors	BD origin	BD (%v/v)	D (%v/v)	B (%v/v)	Year
Atmanli and Yilmaz	-	-	75	25	2018
Atmanli and Yilmaz	-	-	65	35	2018
Kuszewski	-	-	95	5	2018
Kuszewski	-	-	90	10	2018
Kuszewski	-	-	85	15	2018
Kuszewski	-	-	80	20	2018
Kuszewski	-	-	75	25	2018
Álvarez et al.	-	-	95	5	2019
Álvarez et al.	-	-	90	10	2019
Álvarez et al.	-	-	85	15	2019
Álvarez et al.	-	-	80	20	2019
Álvarez et al.	-	-	50	50	2019
Tipanluisa et al.	-	-	95	5	2021
Tipanluisa et al.	-	-	90	10	2021
Tipanluisa et al.	-	-	80	20	2021
Tüccar et al.	Microalgae oil	20	70	10	2014
Tüccar et al.	Microalgae oil	20	60	20	2014
Imbadul et al.	Alexandrian laurel oil	15	70	15	2016a
Imbadul et al.	Alexandrian laurel oil	20	60	20	2016a
Imbadul et al.	Coconut oil	10	80	10	2016b
Imbadul et al.	Coconut oil	20	75	5	2016b
Atmanli	Waste oil	40	40	20	2016
Işık et al.	Not reported	10	80	10	2017
Çelebi and Aydin	Safflower oil	45	45	10	2018
Çelebi and Aydin	Safflower oil	40	40	20	2018

The FP values obtained for all the BD/B and BD/D/B blends studied are shown in Figures 2 and 3. It can be observed that the FP value decreases significantly for low concentrations of B, particularly up to 10% v/v. Thus, two groups of values were considered for the analysis: up to 20% v/v and above 20% v/v of B concentration in the blends. The FP of BD/B blends as a function of the B concentration was measured from 0.5 to 90% v/v of B in BD, whereas the FP of BD/D/B blends was measured from 0 to 90% of B content in the blends, for a constant D concentration of 10% v/v.

The FP values in the binary and ternary systems with B concentration lower than 20% v/v, at a constant temperature, were fitted to a rational function as indicated in equation (1), where x is the B content in the blend (%v/v), Y is the FP of the blend (°C), and A, B, and C are the constants of the fitting.

$$Y = C + \frac{B}{x+A} \qquad (1)$$

The constants of equation (1) and the correlation coefficient (R^2) are listed in Table 2. The maximum difference between experimental and calculated data was 1.3°C (error of 3.5%) for the BD/B blends and 2.5°C (error of 2.3%) for the BD/D/B blends studied. For 20% v/v and higher B concentration in BD, the average FP was 41.8 (+/-0.8)°C, whereas, for B content between 20 and 90% v/v in BD/D, the average FP was 42.4 (+/-1.6)°C.

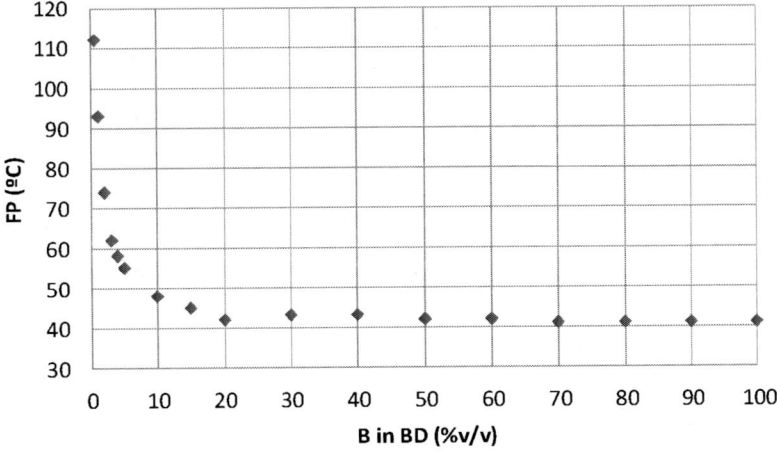

Figure 2. Flash point of biodiesel/butanol blends as a function of butanol concentration.

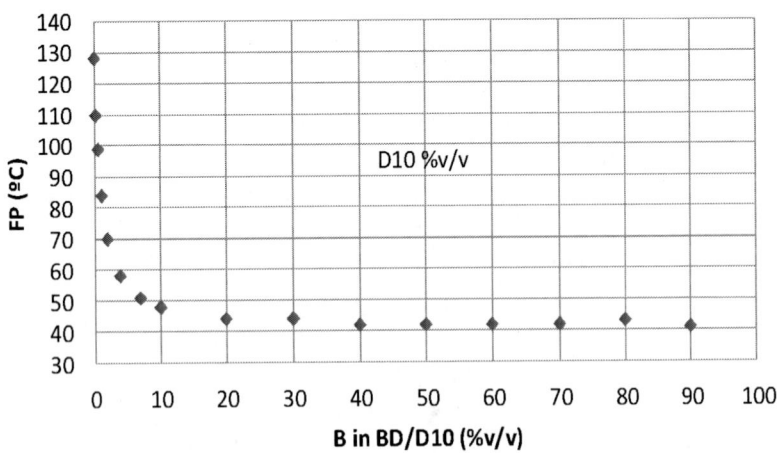

Figure 3. Flash point of biodiesel/diesel/butanol blends as a function of butanol concentration.

Table 2. Fitting constants of equation (2):
Flash point of biodiesel/butanol and biodiesel/diesel/
butanol blends as a function of the butanol concentration

		Constants		
System	A	B	C	R^2
BD/B	0.7177	904.849	389.550	0.998
BD/D/B	0.9597	830.829	408.520	0.998

It is evident that the FP value of the mixtures will depend on the FP values of BD, D, and B used to prepare the blends. In the case of BD, the FP variation range is very wide since it depends - under the same process conditions - on the composition of the fatty acid esters of the starting oil (number of carbons and unsaturation). Although, as a rule, the FP of BD must be higher than 130°C, values can be even higher than 180°C. Consequently, if there were data available in the literature on the variation of FP with the concentration of alcohol in the mixture, only the trends could be compared, unless the reported data correspond to blends with BD produced from the same biomass.

On the other hand, the FP of high-purity B (analytical grade) depends slightly on the commercial brand, and its value varies from 35°C (Imbadul et al., 2016; Yasin et al., 2017; Işık et al., 2017; Atmanli and Yilmaz, 2018; Nanthagopal et al., 2018a; Çelebi and Aydın, 2018) to 41°C (this study).

Finally, the FP of D varies according to the quality or grade and can take different values in different countries. In the literature data revised for this chapter, FP values reported for D are between 58.5 °C (Kuszewski, 2018) and 76.1°C (Campos-Fernández et al., 2012; Atmanli and Yilmaz, 2018), whereas, in this study, the FP value for D was 73°C.

Refractive Index in Biodiesel/Butanol and Biodiesel/Diesel/ Butanol Blends as a Function of Butanol Concentration, at Different Temperatures

The experimental values of RI as a function of the % v/v B content in the BD/B and BD/D/B blends, at different temperatures in the range between 15 and 40°C, were fitted to equation (2)

$$Y = ax + b \qquad (2)$$

where Y is the refractive index, x is the concentration of B in the blend (%v/v), a is the slope (1/%v/v), and b is the Y-intercept.

The RI values for the binary and ternary systems were plotted in Figures 4 and 5. For the sake of clarity, Figure 5 does not include some data at low B concentration, and isotherms at 20, 25, and 35°C. As it can be seen, RI decreases linearly with increasing B concentration, at a constant temperature. The arrow in Figures 4 and 5 goes from lower to higher temperature.

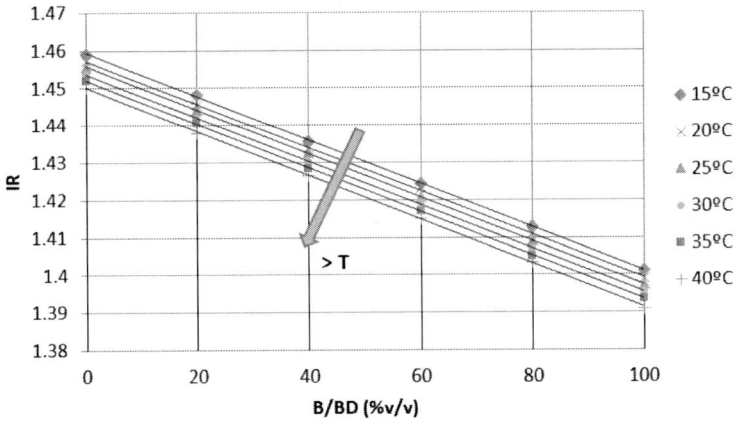

Figure 4. Refractive index as a function of butanol concentration in biodiesel, at different temperatures.

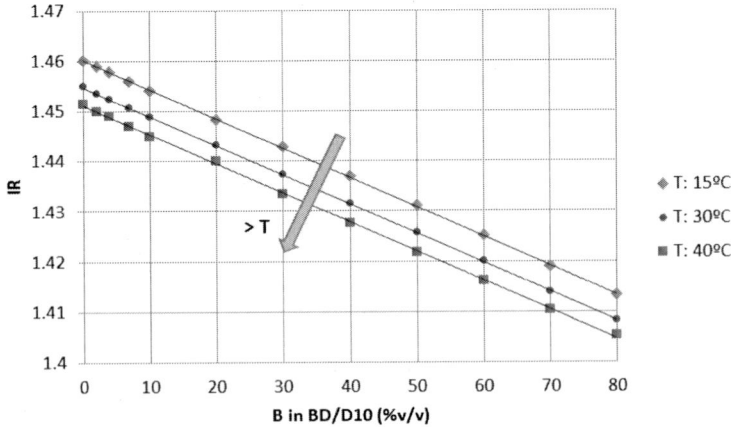

Figure 5. Refractive index as a function of butanol concentration in biodiesel/ diesel blends, at different temperatures.

The RI lines as a function of the B content, in the same system, at different temperature, have similar slopes. The average slope is -5.822 (±0.052) 10^{-4} for the BD/D blends and -5.812 (±0.042) 10^{-4} for BD/D/B blends studied in this work, with a correlation coefficient (R^2) of 0.9998 and 0.9997, respectively.

Refractive Index in Biodiesel/Butanol and Biodiesel/Diesel/ Butanol Blends as a Function of Temperature, at Different Butanol Concentrations

The experimental values of RI as a function of temperature for the BD/B and BD/D/B blends studied are shown in Figures 6 and 7. Figure 6 includes only the RI values of BD, B, and BD/B blends with 20% B increments, and, for the sake of clarity, Figure 7 does not include some data at low B concentration. The arrow in Figures 6 and 7 goes from lower to higher B concentration.

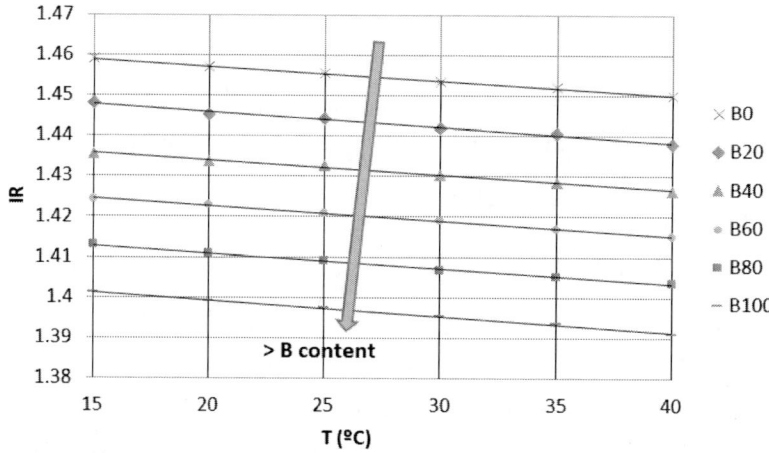

Figure 6. Refractive index as a function of temperature for different concentrations of butanol in biodiesel.

The RI values were adjusted using equation (1). In this case, x is the temperature (°C) for different fixed Bz content, where the z value varies from 0 (pure BD) to 100 (pure B) in Figure 6, and from 0 (90% BD and 10% D) to 80 (80% B, 10% BD, and 10% D) in Figure 7, as indicated in the legend of both Figures. These Figures show that RI decreases linearly with the increase in temperature, and that the slopes are similar for the different compositions in each system. The average value of the slope for the BD/B system is -3.788

(± 0.232) 10^{-4}, whereas that for the BD/D/B blends is -3.609 ($\pm$ 0.179) 10^{-4}. The slope values in BD/D and BD/D/B are similar because the D content in the ternary system is low (10% v/v, constant). R^2 is higher than 0.991 in the binary system and higher than 0.988 in the ternary blends studied.

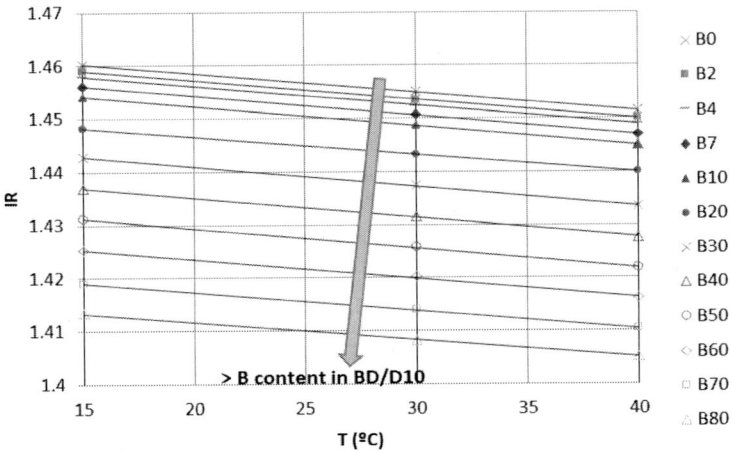

Figure 7. Refractive index as a function of temperature for different concentrations of butanol in biodiesel/diesel blends. Diesel concentration in the blend: 10% v/v.

As mentioned in the introduction section, there are no RI data available in the literature for these systems. However, they show a behavior similar to that of the BD/D blends reported by Colman et al. (2017 and 2018).

Conclusion

The results presented in this chapter provide a deep understanding of the behaviors of the flash point and refractive index (in a temperature range between 15 and 40°C) of the binary BD/B system, in the full concentration range, as well as of a biodiesel/diesel/butanol ternary system with a constant volumetric concentration of diesel (10%).

The flash point values dropped significantly at low butanol concentrations, where a detailed study was carried out, and a very good fitting to a rational function was obtained in both the binary and ternary systems ($R^2 \geq 0.997$). At concentrations of butanol higher than 20% by volume, it can be considered that the flash point remains practically constant.

The refractive index values presented a linear dependence on the butanol concentration and temperature, in both the binary and ternary systems. The higher the butanol content (at a constant temperature) or the higher the temperature (at a constant butanol concentration), the lower the refractive index of the system. The correlation coefficient of the refractive index fitting as a function of butanol concentration at a constant temperature (≥ 0.999) was slightly higher than that as a function of temperature at a constant % v/v butanol content (≥ 0.988). It should be noted that there are no studies in the literature including refractive index measurements as a function of temperature or butanol concentration in these systems.

The results shown in this chapter can be applied to possible quality controls in case of using biodiesel/butanol blends, and as a basis for the study of ternary biodiesel/diesel/butanol systems.

Acknowledgments

This work was supported by the UBACyT projects 20020160100084BA and 20020190100347BA, both from the Universidad de Buenos Aires, and PIDAE 3482, from the Secretaría de Políticas Universitarias, Argentina. The author would like to thank Nicolás Figueroa Semorile, Eng., who made his thesis at GER, for the experimental measurements.

References

Alviso, D., Saab, E., Clevenot, P. and Romano, S. D. (2020). Flash point, kinematic viscosity and refractive index: variations and correlations of biodiesel–diesel blends. *Journal of the Brazilian Society of Mechanical Sciences and Engineering*, 42:347.
Alviso, D. and Romano, S. D. (2021). Prediction of the refractive index and speed of sound of biodiesel from its composition and molecular structure. Fuel, 304,120606.
Altun, S., Öner, C., Yaşar, F. and Adin, H. (2011). Effect of n-butanol blending with a blend of diesel and biodiesel on performance and exhaust emissions of a diesel engine. *Industrial & Engineering Chemistry Research*, 50, 9425-9430.
Atmanli, A. and Yilmaz, N. (2018). A comparative analysis of n-butanol/diesel and 1-pentanol/diesel blends in a compression ignition engine. *Fuel*, 234, 161-169.
Atmanli, A. (2016a). Comparative analyses of diesel–waste oil biodiesel and propanol, n-butanol or 1-pentanol blends in a diesel engine. *Fuel*, 176, 209-215.

Atmanli, A. (2016b). Effects of a cetane improver on fuel properties and engine characteristics of a diesel engine fueled with the blends of diesel, hazelnut oil and higher carbon alcohol. *Fuel*, 172, 209-217.

ASTM D93 International Standard. (2016). Standard test methods for flash point by pensky-martens closed cup tester. *ASTM International*.

ASTM D1218 International Standard. (2016). Standard test method for refractive index and refractive dispersion of hydrocarbon liquids. *ASTM International*.

Babu, M. V., Murthy, K. M. and Rao, G. A. P. (2017). Butanol and pentanol: The promising biofuels for CI engines–A review. *Renewable and Sustainable Energy Reviews*, 78, 1068–1088.

Campos-Fernández, J., Arnal, J. M., Gómez, J. and Dorado, M. P. (2012). A comparison of performance of higher alcohols/diesel fuel blends in a diesel engine. *Applied Energy*, 95, 267–275.

Çelebi, Y. and Aydın H. (2018). Investigation of the effects of butanol addition on safflower biodiesel usage as fuel in a generator diesel engine. *Fuel*, 222, 385-393.

Colman, M., Fernández Galván, E. A., Sorichetti, P. A., and Romano, S. D. (2017). Estudio de mezclas diesel – biodiesel por refractometría en el rango visible. Aplicación al corte argentino [*Study of diesel -biodiesel mixtures by refractometry in the visible range. Application to the Argentine court*]. *Anales AFA*, 28 (1), 1-3.

Colman, M., Sorichetti, P. A., and Romano, S. D. (2018). Refractive index of biodiesel-diesel blends from effective polarizability and density. *Fuel*, 211, 130-139.

Geacai, E., Nita, I., Osman, S. and Iulian, O. (2017). Effect of n-butanol addition on density and viscosity of biodiesel, *U. P. B. Sci. Bulletin*, Series B, 79, 1, 11–24.

Imdadul, H. K., Masjuki, H. H., Kalam, M. A., Zulkifli, N. W. M., Alabdulkarem, A. Kamruzzaman, M., Rashed, M. M. (2016a). A comparative study of C4 and C5 alcohol treated diesel–biodiesel blends in terms of diesel engine performance and exhaust emission. *Fuel*, 179, 281-288.

Imdadul, H. K., Masjuki, H. H., Kalam, M. A., Zulkifli, N. W. M., Alabdulkarem, A., Rashed, M. M. and Ashraful, A. M. (2016). Influences of ignition improver additive on ternary (diesel-biodiesel-higher alcohol) blends thermal stability and diesel engine performance. *Energy Conversion and Management*, 123, 252-264.

Imdadul, H. K., Masjuki, H. H., Kalam, M. A., Zulkifli, N. W. M., Alabdulkarem, A., Rashed, M. M., Teoh, Y. H. and Howa, H. G. (2016). Higher alcohol–biodiesel–diesel blends: An approach for improving the performance, emission, and combustion of a light-duty diesel engine. *Energy Conversion and Management*, 111, 174-185.

Işık, M. Z., Bayındır, H., Iscan, B. and Aydın, H. (2017). The effect of n-butanol additive on low load combustion, performance and emissions of biodiesel-diesel blend in a heavy duty diesel power generator. *Journal of the Energy Institute*, 90, 174-184.

Knothe, G. (2005). Dependence of biodiesel fuel properties on the structure of fatty acid alkyl esters. *Fuel Processing Technology*, 86, 1059-1070.

Knothe, G., Van Gerpen, J. and Krahl, J. (2005). *The Biodiesel Handbook*. AOCS Press.

Kuszewski, H. (2018). Physical and chemical properties of 1-butanol–diesel fuel blends. *Energy & Fuels*, 32, 11, 11619–11631.

Lapuerta, M., García–Contreras, R., Campos–Fernández, J. and Dorado, M. P. (2010). Stability, lubricity, viscosity, and cold-flow properties of alcohol-diesel blends. *Energy & Fuels*, 24, 4497-4502.

Lapuerta, M., Rodríguez-Fernández, J., Fernández-Rodríguez, D. and Patiño-Camino, R. (2018). Cold flow and filterability properties of n-butanol and ethanol blends with diesel and biodiesel fuels. *Fuel*, 224, 552-559.

Lebedevas, S., Lebedeva, G., Sendzikiene, E. and Makareviciene, V. (2010). Investigation of the performance and emission characteristics of biodiesel fuel containing butanol under the conditions of diesel engine operation. *Energy & Fuels*, 24, 4503–4509.

Nanthagopal, K., Ashok, B., Saravanan, B., Korah, S. M. and Chandra, S. (2018b). Effect of next generation higher alcohols and Calophyllum inophyllum methyl ester blends in diesel engine. *Journal of Cleaner Production*, 180, 50-63.

Nanthagopala, K., Ashoka, B., Saravananb, B., Patela, D., Sudarshana, B. and Ramasamya, R. A. (2018a). An assessment on the effects of 1-pentanol and 1-butanol as additives with Calophyllum Inophyllum biodiesel. *Energy Conversion and Management*, 158, 70-80.

Romano, S. D., Gozález Suárez, E. and Laborde, M. A. (2005). Biodiesel. *Combustibles alternativos*. Ediciones Cooperativas. Capital Federal, Argentina.

Siwale, L., Siwale, L., Kristóf, L., Adam, T., Bereczky, A., Mbarawa, M., Penninger, A. and Kolesnikov, A. (2013). Combustion and emission characteristics of n-butanol/diesel fuel blend in a turbo-charged compression ignition engine. *Fuel*, 107, 409-418.

Tipanluisa, L., Fonseca, N., Casanova, J. and López, J. M. (2021). Effect of n-butanol/diesel blends on performance and emissions of a heavy-duty diesel engine tested under the World Harmonised Steady-State cycle. *Fuel*, 302, 121204.

Tüccar, G., Özgür, T., and Aydın, K. (2014). Effect of diesel – microalgae biodiesel – butanol blends on performance and emissions of diesel engine. *Fuel*, 132, 47-52.

Valentino, G., Corcione, F. E. and Iannuzzi, S. E. (2012). Effects of gasoline–diesel and n-butanol–diesel blends on performance and emissions of an automotive direct-injection diesel engine. *International Journal of Engine Research*, 13 (3), 199-215.

Yang, M., Wang, Z., Guo, S., Xin, X., Qi, T., Lei, T. and Yan X. (2017). Effects of fuel properties on combustion and emissions of a direct injection diesel engine fueled with n-butanol-diesel blends. *Journal of Renewable and Sustainable Energy*, 9, 1, 013105.

Yasin, M. H. M., Mamat, R., Yusop, A. F., Abdullah, A. A., Othman, M. F., Yusrizal, S. T. and Iqbal S. T. (2017). Cylinder pressure cyclic variations in a diesel engine operating with biodiesel-alcohol blends. *Energy Procedia*, 142, 303-308.

Yilmaz, N. and Vigil, F. M. (2014). Potential use of a blend of diesel, biodiesel, alcohols and vegetable oil in compression ignition engines. *Fuel*, 124,168–172.

Yoshimoto, Y., Kinoshita, E., Shanbu, L., Ohmura, T. (2013). Influence of 1-butanol addition on diesel combustion with palm oil methyl ester/gas oil blends. *Energy*, 61, 44-51.

Wei, L., Cheung, C. S., Huang, Z. (2014). Effect of n-pentanol addition on the combustion, performance and emission characteristics of a direct-injection diesel engine. *Energy*, 70, 172-180.

Chapter 5

Natural Attenuation of Soil with Biodiesel and Its Potential Ecotoxicological Impacts

Guilherme Dilarri[1], Carolina Rosai Mendes[1], Vinícius de Moraes Ruy Sapata[1], Ivo Shodji Tamada[1], Paulo Renato Matos Lopes[2], Renato Nallin Montagnolli[3] and Ederio Dino Bidoia[1,1]

[1]Department of General and Applied Biology,
Sao Paulo State University (UNESP), Rio Claro-SP, Brazil
[2]Department of Plant Production,
College of Agricultural and Technological Sciences,
Sao Paulo State University (UNESP), Dracena-SP, Brazil
[3]Department of Natural Sciences, Mathematics and Education,
Agricultural Sciences Centre,
Federal University of Sao Carlos (UFSCar), Araras-SP, Brazil

Abstract

Petroleum and its derivatives cause numerous toxicity-related impacts on the environment. The evaluation of such impacts can promote adequate treatments of oil-based residues. A better understanding of the degradation of hydrocarbons aids researchers to predict the environmental impacts and the biodegradation processes that yield decreased toxicity levels. However, petroleum derivatives are not the sole matter of concern. Another important discussion targeting the environmental behavior of biodiesel should be brought up. Soil microorganisms may improve the biodegradation of hydrocarbons due to their diversity and robust enzymatic apparatus during natural attenuation

---

[1] Corresponding Author's Email: ederio.bidoia@unesp.br.

In: The Future of Biodiesel
Editor: Michael F. Simpson
ISBN: 979-8-88697-166-8
© 2022 Nova Science Publishers, Inc.

processes. Still, some compounds can have their negative impacts on biota evaluated according to three parameters: ecotoxicological potential in soil (i), specific biodegradation processes (ii), and post-treatment bioassays (iii). This review presents how microbial biodegradation may ultimately affect the overall degradation of hydrocarbons.

Keywords: bioassays, bioremediation, ecotoxicity, petroleum, fuels

Analytical Literature Review

Petroleum

Petroleum has been the dominant energy source since the last century due to its central role in the modernization and the development of our society. Most of it is based on the heavy reliance of automobile and industrial machinery sectors on petroleum refined derivatives - gasoline, fuel oil, heating oil, and other products (Ederington et al., 2019). Furthermore, a wide range of products needs petroleum to a certain extent in their manufacturing process. Although their large-scale production revolutionized the industry of materials science and engineering - plastics, electronics, and novel construction parts (Saleem et al., 2018) - such production also has its environmental drawbacks.

Crude oil is not man-made. It naturally occurs underneath the Earth's surface, whose generation depends on a large volume of organic matter accumulated during the formation of sedimentary rocks under very specific conditions. The layers of source rocks (mostly shales and lime mudstones) generate petroleum when subject to certain temperatures and pressures in oxygen-free pockets of porous and permeable rocks (Milani et al., 2000). The final composition depends on the conditions and the organic matter available from the fossilized sources: zooplankton and phytoplankton often produce a higher oil ratio whereas vegetable sources often yield a higher natural gas ratio. Moreover, the process output in the source rock is temperature-dependent, as values below 600°C generate mostly oil, and values above 1200°C produce mostly natural gas (Milani et al., 2000).

The resulting hydrocarbon mixture is extremely complex and may contain variable amounts of sulfur, oxygen, and nitrogen-rich organic compounds (Ossai et al., 2020). These molecules include linear and branched alkanes chains, cycloalkanes, unsaturated alkenes, alkynes, aromatics, and polycyclic aromatic hydrocarbons (PAHs) such as naphthalene or monoaromatic compounds (benzene, toluene, ethylbenzene, and xylene - BTEX), asphaltenes, phenols, esters, porphyrins, fatty acids. The heavier compounds include resins, waxes, and tars. Alcohols, carboxylic acids, ethers, ketones, furans, pyrrole, pyridine, carbazole, nitriles, indoline, quinoline, sulfides, thiols, cyclic sulfides, disulfides, dibenzothiophene, benzothiophene and naphthobenzothiphene (Ossai et al., 2020) can also be expected in the final crude petroleum mixture. Thus, the contamination by petroleum and its derivatives could seriously harm the environment based on our current understanding of the high toxicity of many of these molecules.

Biodiesel

Biodiesel has its share in fulfilling an ever-increasing global energy demand in a sustainable way. Biodiesel is derived from renewable sources, such as vegetable oils from soybean, sunflower, and peanut (Rico and Sauer, 2015).

Biodiesel is produced from esters via the methyl route. Methanol is the main raw material for obtaining methyl esters. Methanol has high toxicity and is extracted from non-renewable fossil sources; however, methanol may also be obtained by distilling cellulose (Hassan & Kalam, 2013). Ethanol has similar fuel and energy properties to methanol, with the advantage of being less toxic (Meyer et al., 2014).

Compared to petroleum-derived fuel, biodiesel usage reduces carbon dioxide emissions by 80% and sulfur oxide emissions by 90% (Hawrot-Paw et al., 2019). Biodiesel can easily replace conventional diesel directly into diesel cycle engines, without any adaptation or mechanical modifications. In addition, biodiesel represents an important breakthrough for energy self-sufficiency and strategic economic sectors targeting the export of commodities.

Biodiesel, despite being an organic compound, has its formulation altered after biodegradation which could lead to toxicity and several environmental impacts (Montagnolli et al., 2019). Montagnolli et al. (2019) reported that the ecotoxicological effects from the byproducts of biodiesel's biodegradation can be mitigated using biosurfactants, which raised their solubility by up to 31%. In addition, the biodegradation rate increased due to the availability of biodiesel to the microbial populations. Therefore, these are potential strategies for short-term bioremediation and natural attenuation that successfully remove biodiesel contamination from the environment.

According to studies by Cruz et al. (2019), biodiesel produced from animal fat has intrinsic characteristics that differ from those found in vegetable-based biodiesel. Their research group performed bioassays after the biodegradation with indigenous soil microbiota which showed that biodiesel from animal fats generated less toxic biodegradation byproducts compared to the vegetable-based ones. Despite biodiesel being a carbon source for several microorganisms, there are mixed, and often contradictory results, concerning its predicted environmental impacts. These results may be biased by narrow comparisons between biofuels and fossil fuels. Still, the consensus is that biodiesel can be bioaccumulative in the environment, thus impacting the soil microbial biodiversity.

Environmental Impacts

The unrestrained use of petroleum and its derivatives are affecting the global carbon cycle, thus accelerating the greenhouse effect and climate change. Besides, many potential environmental problems are aggravated by the logistics of the petroleum industry during accidents involving refineries, tankers, spills, and seepages. Furthermore, our society has yet to figure out ways to effectively discard the increasing amounts of solid plastic waste (Saleem et al., 2018). These situations cause severe consequences to local biota considering their carcinogenic, toxicity, teratogenic and mutagenic effects (Wang et al., 2017; Ahmed & Fakhruddin, 2018; Koshlaf et al., 2020).

The complex composition of hydrocarbons makes it hard to effectively decrease or completely neutralize petroleum pollution (Koshlaf et al., 2020).

A single solution may not be the answer. Whereas no universal solution can be proposed, it's been reported that a combination of two or more methods (or processes) could work much better. Especially considering that some compounds are hydrophobic and present a high fusion and boiling point, which provides an additional level of difficulty to remediate contaminated areas (Wang et al., 2017).

Biodiesel contamination may also directly affect the local biota, yet its toxicity is much lower compared to fossil fuels. Since biodiesel is a carbon source for various soil microorganisms, some positive results from microbial community analyses are expected when investigating the overall environmental impact. Still, it is imperative to consider the post-biodegradation effects regarding toxicity. Hawrot-Paw et al. (2019) showed promising results in their biodiesel biodegradation assays after phytostimulation with *Pisum sativum* in agricultural soil. However, other authors indicated that biofuel could be hazardous and induce many long-term adverse environmental impacts. Pikula et al. (2020) showed genotoxic and mutagenic effects in marine microalgae communities exposed to biodiesel. Leme et al. (2012) observed a lower germination rate in several plant species, as well as lower microbial biomass production in soils containing biodiesel.

Physical methods are also an alternative for remediating soils. Incineration, soil vapor extraction, and soil washing are often proposed. Nevertheless, these technologies are either expensive or highly energy-demanding. Physical removal is often tied to secondary problems (Ahmed & Fakhruddin, 2018).

Chemical treatments exist as well. They consist of abiotic transformations by degrading chemicals through hydrolysis, photolysis, oxidation-reduction, and/or mineralization. Still, these are not cost-effective techniques either and are seldom recommended as a complementary remediation method (Ahmed & Fakhruddin, 2018; Ossai et al., 2020).

Bioremediation

There is a demand for cheaper technologies to deal with such impacted environments. Several advances in bioremediation research have been developed through many biological agents. Microbial communities that are naturally present in polluted areas showed a high variety of catabolic pathways to biodegrade hydrocarbons via oxidation (Xu et al., 2018).

Several studies have isolated bacterial strains from oil-contaminated environments that are capable of degrading hydrocarbons: *Acinetobacter, Burkholderia, Enterobacter, Rhodococcus, Pandoreae, Kocuria Mycobacterium, Pseudomonas* (Margesin et al., 2003; Sarkar et al., 2017; Xu et al., 2018). The complete set of environmental parameters also influences biodegradability: nutrients availability, soil type, aeration, pH, temperature, porosity, etc. The target compound has its biodegradability rates linked to concentration, toxicity, hydrophobicity, etc. (Koshlaf et al., 2020; Wang et al., 2017).

Although biodegradation optimization emerges from increasing environmental awareness and the search for eco-friendly methods for bioremediation, they still take a long time to be effective in practical scenarios (Ahmed & Fakhruddin, 2018; Xu et al., 2018). The reasoning for bio-remediation studies aims to increase the applicability of these methods by enhancing microbial biodegradation output. Ossai et al. (2020) and Xu et al. (2018) argue that many strategies can be proposed to enhance metabolic rates within the soil microbiota by adding oxygen (bioaeration), enhanced bacterial and fungal strains (bioaugmentation), nutrients (biostimulation), and surfactants (biosolubilization).

Regardless of the chosen remediation protocol applied to a polluted site, it is essential to continuously monitor the environmental conditions. Ideally, every biodegradation process should require a comparison between the initial and the final state. Ecotoxicological data provide a reliable success measurement for biodegradation processes by using adequate bioindicator species.

Ecotoxicity

Ecotoxicology focuses on evaluating the environmental impacts caused by anthropogenic actions through the analysis of deviant patterns in mortality, reproduction, growth, and metabolism of bioindicator species. These bioassays allow proper ecological risk assessment and are validated by robust statistical tools (Kobetičová & Černý, 2017). This last step in bioremediation dataset analysis is important to demonstrate an effective treatment.

The complete degradation of organic contaminants by microbial metabolism should optimally yield only two final products: CO_2 and H_2O. However, some biochemical reactions may generate intermediate compounds whose toxicity could be higher than the original contaminant. Therefore, even though biodegradation occurs, it does not necessarily mean actual bioremediation (Ahmed & Fakhruddin, 2018; Koshlaf et al., 2020; Ossai et al., 2020).

Ecotoxicity tests employ organisms that are sensitive to subtle environmental changes as well as vulnerable to contaminants even at extremely low concentrations. Vegetables (*Allium cepa L.*, *Lactuca sativa*, *Eruca sativa*), soil meso- and microfauna (*Eisenia foetida*, *Eisenia andreii*, *Enchytraeus crypticus*, *Folsomia candida*), and microcrustacean such as *Daphnia magna* and *Ceriodaphnia dubia* (França et al., 2021) are often chosen for ecotoxicity bioassays.

Tamada et al. (2012) used three organisms for toxicity tests: *Eruca sativa* seeds (arugula), *Lactuca sativa* seeds (lettuce), and *Eisenia andrei* (earthworms). Their bioassays using seeds of *E. sativa* showed a decrease in the toxicity levels of both mineral and synthetic lubricant oils (both petroleum derivatives) after biodegradation. *E. sativa*, exposed to the oils before biodegradation, however, had their germination inhibited in more than 80% of the subjects. Interestingly, biodegradation assays containing mineral oil showed a toxicity increase throughout the first 60 days before finally decreasing its toxicity at the end of the mineralization process. These results are linked to the expected mid-point formation of toxic biodegradation byproducts.

A research report by Tamada et al. (2012) using 10.0 mL of lubricating oil per 100 g of soil showed that mortality of *E. andrei, L. sativa*, and *E. sativa* successfully decreased when such oils were biodegraded. Biodegradation proved to be effective for mineral and synthetic oils according to their bioassays.

Bioassays are, therefore, encouraged when validating waste treatment or bioremediation strategies. The choice of an adequate test organism can widely expand current knowledge on the safe biodegradation of any compound regardless of the many widely available bioremediation procedures.

Conclusion

Biodegradation is more effective in biodiesel than in petroleum-based fuels. Biodiesel is indeed much easier to biodegrade; however, its composition can change according to the raw material used. Depending on the source, the complexity of the molecules interferes with the microbiological treatment time. Still, biodiesel is a welcome alternative to fossil fuels, as it reduces carbon dioxide emissions during combustion. Most studies indicated reduced toxic effects of biodiesel after biodegradation. Its toxicity, however, is not a consensus due to mixed experimental results widely debated in current references.

Thus, bioassays are required to assess environmental impacts and ecological risks. Sensitive test organisms are important to verify the decrease in toxicity after microbiological treatments, or simply when monitoring natural attenuation. Biodiesel contamination directly affects biota, but most studies have indicated low toxicity levels after biodegradation and consequently less impact on the environment than fossil fuels.

References

Ahmed, F., & Fakhruddin, A. N. M. (2018) A review on environmental contamination of petroleum hydrocarbons and its biodegradation. *International Journal of Environmental Sciences & Natural Resources*, 11, 1-7.

Cruz, J. M., Corroqué, N. A., Montagnolli, R. N., Lopes, P. R. M., Morales, M. A. M., Bidoia, E. D. (2019) Comparative study of phytotoxicity and genotoxicity of soil contaminated with biodiesel, diesel fuel and petroleum. *Ecotoxicology*, 28, 449-456.

Ederington, L. H., Fernando, C. S., Hoelscher, S. A., Lee, T. K., Linn, S. C. (2019) A review of the evidence on the relation between crude oil prices and petroleum product prices. *Journal of Commodity Markets*, 13, 1-15.

França, M., Emmerich, M., Oliveira, T. M. N., Franczack, P., Oliveira M. B. G. (2021) Toxicity study of the sanitary wastewater treatment plant sludge aiming its reuse in agriculture. *Water, Air, & Soil Pollution*, 232, 1-12.

Hassan, M. H. & Kalam, M. A. (2013) An overview of biofuel as a renewable energy source: development and challenges. *Procedia Engineering*, 56, 39-53.

Hawrot-Paw, M., Ratomski, P., Mikiciuk, M., Staniewski, J., Koniuszy, A., Ptak, P., Golimowski, W. (2019) Pea cultivar Blauwschokker for the phytostimulation of biodiesel degradation in agricultural soil. *Environmental Science and Pollution Research*, 26, 34594-34602.

Kobetičová, K., & Černý, R. (2017) Ecotoxicology of building materials: A critical review of recent studies. *Journal of Cleaner Production*, 165, 500-508.

Koshlaf, E., Shahsavari, E., Haleyur, N., Osborn, A. M., Ball, A. S. (2020) Impact of necrophytoremediation on petroleum hydrocarbon degradation, ecotoxicity and soil bacterial community composition in diesel-contaminated soil. *Environmental Science and Pollution Research*, 27, 31171-31183.

Leme, D. M., Grummt, T., Heinze, R., Sehr, A., Renz, S., Reinel, S., Palma de Oliveira, D., Ferraz, E. R. A., Rodrigues de Marchi, M. R., Machado, M. C., Zocolo, G. J., Marin-Morales, M. A. (2012) An overview of biodiesel soil pollution: data based on cytotoxicity and genotoxicity assessments. *Journal of Hazardous Materials*, 200, 343-349.

Margesin, R., Labbe, D., Schinner, F., Greer, C. W. & Whyte, L. G. (2003) Characterization of hydrocarbon-degrading microbial populations in contaminated and pristine alpine soils. *Applied and Environmental Microbiology*, 69, 3085-3092.

Meyer D. D., Beker S. A., Bucker F., Peralba M. C. R., Frazzon A. P. G., Osti J., Andreazza R., Camargo F. A. O., Bento F. M. (2014) Bioremediation strategies for diesel and biodiesel in oxisol from southern Brazil. *International Biodeterioration & Biodegradation*, 95, 356-363.

Milani, E. J.; Brandão, J. A. S. L., Zalán, P. V. & Gamboa, L. A. P. (2000) Petróleo na margem continental brasileira: geologia, exploração, resultados e perspectivas [*Oil on the Brazilian continental margin: geology, exploration, results and prospects*]. *Revista Brasileira de Geofísica*, 18, 352-396.

Montagnolli, R. N., Cruz, J. M., Moraes, J. R., Mendes, C. R., Dilarri, G., Lopes, P. R. M., & Bidoia, E. D. (2019). Technical approaches to evaluate the surfactant-enhanced biodegradation of biodiesel and vegetable oils. *Environmental Monitoring and Assessment*, 191, 565.

Ossai, I. C.; Ahmed, A.; Hassan, A. Hamid, F. S. (2020) Remediation of soil and water contaminated with petroleum hydrocarbon: A review. *Environmental Technology & Innovation*, 17, 100526.

Pikula, K.; Zakharenko, A.; Chaika, V.; Kirichenko, K.; Tsatsakis, A.; Golokhvast, K. (2020) Risk assessments in nanotoxicology: Bioinformatics and computational approaches. *Current Opinion in Toxicology*, 19, 1-6.

Rico, J. A. P., & Sauer, I. L. (2015). A review of Brazilian biodiesel experiences. *Renewable & Sustainable Energy Reviews*, 45, 513-529.

Saleem, J., Riaz, M. A., Gordon, M. (2018) Oil sorbents from plastic wastes and polymers: A review. *Journal of Hazardous Materials*, 341, 424-437.

Sarkar, P., Roy, A., Pal, S., Mohapatra, B., Kazy, S. K., Maiti, M. K. Sar, P. (2017) Enrichment and characterization of hydrocarbon-degrading bacteria from petroleum refinery waste as potent bioaugmentation agent for in situ bioremediation. *Bioresource Technology*, 242, 15-27.

Tamada, I. S., Lopes, P. R. M., Montagnolli, R. N., Bidoia, E. D. (2012) Biodegradation and toxological evaluation of lubricant oils. *Brazilian Archives of Biology and Technology*, 55, 951-956.

Xu, X., Liu, W., Tian, S., Wang, W., Qi, Q., Jiang, P., Gao, X., Li, F., Li, H., Yu, H. (2018) Petroleum hydrocarbon-degrading bacteria for the remediation of oil pollution under aerobic conditions: a perspective analysis. *Frontiers in Microbiology*, 9, 2885.

Wang, S.; Xu, Y.; Lin, Z.; Zhang, J.; Norbu, N., Liu, W. (2017) The harm of petroleum-polluted soil and its remediation research. In: *AIP Conference Proceedings*. AIP Publishing LLC, 020222.

Index

A

acid, 15, 17, 19, 20, 21, 28, 31, 33, 34, 39, 43, 45, 48, 49, 51, 54, 59, 61, 63, 64, 65, 69, 89, 90, 93, 95, 96, 98, 101, 107, 142, 149
agriculture, 101, 105, 162
alcohols, 16, 31, 102, 126, 134, 135, 148, 150
algae, 2, 4, 6, 16, 23, 26, 27, 42, 49, 50
anaerobic sludge, 15, 16, 29, 45
aquaculture, 18, 38, 39

B

bacteria, 5, 6, 28, 37, 40, 41, 42, 45, 48, 49, 50, 163, 164
bioassays, xii, 154, 156, 160, 161
biodegradation, ix, xii, 4, 154, 156, 158, 159, 160, 161, 162, 163
bioethanol, vii, ix, xi, 33, 53, 99, 100, 101, 105, 109, 112, 113, 114, 115, 116, 117, 119, 120, 121, 122, 123, 124, 125, 126, 127, 128, 129, 130, 131
biofuel, ix, 1, 8, 38, 41, 43, 44, 47, 50, 101, 102, 127, 134, 135, 158, 162
biomass, ix, 1, 3, 4, 6, 7, 8, 16, 17, 20, 21, 22, 23, 24, 26, 27, 29, 30, 31, 32, 33, 34, 35, 36, 38, 39, 41, 44, 45, 46, 47, 48, 49, 50, 51, 52, 53, 54, 55, 130, 134, 142, 158
bioremediation, 154, 156, 158, 159, 160, 161, 163
blends, vii, ix, xi, xii, 104, 105, 112, 116, 117, 119, 120, 121, 122, 123, 125, 126, 127, 128, 129, 130, 131, 133, 134, 135, 136, 137, 138, 139, 140, 141, 142, 143, 144, 145, 146, 147, 148, 149, 150, 151
butanol, vii, ix, xi, xii, 34, 104, 131, 133, 134, 135, 136, 137, 138, 139, 141, 142, 143, 144, 145, 146, 147, 148, 149, 150, 151

C

carbohydrates, 3, 16, 32
carbon, 2, 8, 37, 50, 90, 120, 122, 135, 148, 156, 157, 158, 161
carbon dioxide, 50, 122, 156, 161
carbon monoxide, 120
carboxylic acids, 155
castor oil, 60, 63, 64, 65
catalysis, 67, 82, 91, 92
catalyst, 19, 32, 33, 35, 59, 61, 64, 66, 67, 68, 69, 89, 90, 91, 92, 93, 94, 95, 97, 98, 107
cellulose, 27, 53, 156
chemical, x, 3, 21, 25, 26, 49, 55, 57, 76, 81, 83, 87, 88, 91, 94, 96, 98, 102, 149
chemicals, 25, 27, 49, 158
chitosan, 27, 36, 92
chloroform, 30, 31, 35
climate, 59, 101, 157
climate change, 59, 101, 157
CO_2, xi, 2, 5, 6, 18, 19, 30, 35, 41, 50, 59, 84, 100, 104, 108, 122, 127, 160
combustion, xi, 95, 101, 103, 110, 112, 113, 115, 117, 119, 120, 121, 122, 123, 124, 125, 126, 128, 129, 130, 131, 133, 136, 149, 150, 151, 161
compounds, xii, 2, 27, 68, 84, 102, 154, 155, 157, 160

Computational Fluid Dynamics, 58, 60, 70, 90, 91, 94, 95
consumption, xi, 6, 26, 30, 31, 76, 80, 82, 100, 101, 104, 105, 109, 111, 117, 127, 129, 133
contamination, 155, 156, 157, 161, 162
cultivation, 9, 10, 12, 14, 15, 20, 36, 37, 38, 39, 41, 42, 45, 47, 48, 52, 53, 54, 128
culture, 8, 14, 20, 40, 41, 44, 48, 53
cylinder pressure, xi, 100, 107, 108, 110, 117, 120, 126, 150
cynara cardunculus, xi, 99, 100, 101, 105, 107, 127, 128, 130

D

diesel, vii, ix, xi, xii, 95, 99, 100, 101, 102, 103, 104, 105, 106, 109, 111, 113, 114, 115, 116, 117, 118, 119, 120, 121, 122, 123, 124, 125, 126, 127, 128, 129, 130, 131, 133, 134, 135, 136, 137, 138, 139, 142, 143, 144, 145, 146, 147, 148, 149, 150, 151, 156, 162, 163

E

ecotoxicity, 154, 160, 162
effluent, ix, 1, 3, 5, 6, 8, 11, 12, 13, 14, 18, 19, 20, 21, 25, 29, 37, 40, 48, 53
emission, xi, 49, 58, 100, 101, 104, 107, 108, 109, 120, 122, 123, 124, 125, 126, 127, 128, 129, 130, 131, 149, 150, 151
energy, ii, ix, x, 1, 2, 14, 16, 24, 25, 26, 28, 30, 31, 32, 34, 35, 44, 51, 58, 69, 71, 82, 88, 101, 104, 115, 126, 129, 130, 134, 154, 156, 158, 162
environment, xii, 4, 5, 43, 110, 153, 155, 156, 157, 162
environmental impact, xii, 14, 153, 156, 157, 158, 160, 161
ethanol, 33, 34, 35, 60, 61, 63, 64, 70, 87, 96, 102, 103, 104, 105, 116, 127, 128, 129, 130, 131, 149

F

fatty acids, 16, 20, 32, 35, 82, 88, 96, 155
feedstock, 20, 31, 45, 51, 61, 63, 66, 68
flash point, ix, xii, 102, 103, 105, 106, 133, 134, 136, 138, 141, 142, 146, 147, 148
flocculation, 23, 25, 26, 27, 28, 29, 43, 44, 46, 50, 52, 53, 55
fluid, 30, 35, 60, 70, 71, 72, 73, 74, 75, 76, 78, 88, 91, 93, 96
food, x, 31, 58, 101
fouling, 6, 25, 54
freshwater, 27, 47
fuel consumption, xi, 100, 101, 104, 105, 109, 111, 117, 129, 133
fuels, 58, 80, 90, 91, 101, 103, 104, 109, 116, 126, 127, 129, 134, 135, 149, 150, 154, 157, 158, 161, 162
fungi, 5, 20, 29, 37, 38

G

global warming, 14, 35, 59
glucose, 39, 63, 93, 98
glycerol, 31, 59, 67, 68, 69, 90, 92
glycine, 29, 54
green alga, 23, 26, 27, 28, 38, 50, 51
greenhouse, 2, 49, 157
growth, ix, 1, 2, 4, 5, 7, 8, 9, 12, 13, 14, 15, 20, 21, 23, 29, 38, 40, 47, 51, 160

H

harvesting, x, 2, 4, 6, 8, 21, 22, 23, 25, 26, 27, 28, 29, 36, 37, 38, 39, 41, 43, 44, 45, 46, 47, 48, 50, 51, 52, 53, 54, 55
health, 44, 136
heat release, xi, 100, 117, 119, 120, 126, 127, 129
heat transfer, 80, 83, 84, 110
heterogeneous catalysis, 67, 92
hexane, 19, 30, 31, 34, 35
HRR, xi, 100, 109, 118, 119, 120, 125
hydrocarbons, ix, xii, 104, 135, 153, 157, 159, 162
hydrogen, 37, 86, 129, 130, 131

Index

I

industry, 7, 33, 83, 101, 102, 154, 157

K

kinetics, 60, 74, 76, 90

L

laminar, 9, 60, 63, 64, 75, 119
laws, 70, 72, 78
lead, 2, 6, 124, 126, 156
light, 2, 3, 4, 5, 6, 9, 10, 11, 18, 20, 21, 51, 138, 149
lipids, 3, 16, 30, 31, 33, 36, 38, 39, 49, 54

M

machinery, 134, 154
macronutrients, 2, 21
magnetic field, 15, 25, 26, 39
materials, 24, 68, 93, 98, 103, 105, 154, 162
matrixes, 85, 86, 89
metabolism, 5, 20, 21, 160
methanol, 19, 30, 33, 34, 35, 36, 49, 59, 61, 63, 66, 67, 81, 88, 90, 92, 93, 95, 96, 102, 104, 105, 107, 131, 156
microalgae, vii, ix, 1, 2, 3, 4, 5, 6, 8, 14, 16, 20, 22, 24, 25, 28, 29, 30, 31, 34, 36, 37, 38, 39, 40, 41, 42, 43, 44, 45, 46, 47, 48, 49, 50, 51, 52, 53, 54, 55, 140, 150, 158
microorganisms, xii, 48, 154, 157, 158
microreactor, 58, 59, 61, 63, 64, 66, 68, 69, 74, 80, 81, 82, 83, 84, 86, 87, 89, 90, 91, 92, 93, 94, 95, 96, 97
mineralization, 158, 161
missions, xi, 100, 103, 104, 127, 129, 156, 161
mixing, x, 5, 9, 21, 26, 38, 58, 60, 61, 63, 64, 65, 67, 74, 81, 88, 93, 96, 97, 121, 124
molecules, 2, 12, 21, 24, 124, 155, 161

N

Na^+, 92
NaCl, 21
nanoparticles, 25, 36
naphthalene, 155
nitrogen, 38, 39, 40, 127, 155
non-polar, 16, 30, 31
nutrients, ix, 1, 3, 4, 5, 6, 9, 14, 17, 21, 29, 32, 38, 39, 40, 41, 45, 46, 50, 51, 54, 159

O

oil, x, xi, xii, 2, 11, 18, 29, 30, 31, 35, 37, 53, 58, 60, 61, 63, 64, 65, 66, 67, 68, 69, 80, 81, 87, 88, 89, 90, 91, 92, 93, 94, 95, 96, 97, 100, 102, 105, 107, 127, 128, 130, 134, 139, 140, 142, 148, 151, 153, 154, 155, 159, 160, 161, 162, 163, 164
organic compounds, 2, 84, 155
organic matter, 25, 37, 41, 50, 52, 155
oscillation, 63, 93
osmosis, 24, 46, 48
oxidation, 36, 38, 84, 103, 136, 158, 159
oxygen, 3, 40, 102, 103, 112, 113, 115, 119, 120, 121, 122, 124, 125, 126, 134, 155, 159
ozone, 24, 42

P

palm oil, 18, 37, 81, 90, 151
petroleum, xii, 30, 134, 153, 154, 155, 156, 157, 160, 161, 162, 163, 164
pH, 4, 26, 27, 38, 39, 50, 159
pharmaceuticals, x, 4, 40, 58, 83
phycoremediation, ix, 1, 2, 48
phytoremediation, 2, 3, 37
plants, 70, 83, 89, 101
polar, 16, 30, 31
pollutants, 4, 104, 129
pollution, 54, 101, 135, 157, 163, 164
polyacrylamide, 28, 37, 47
polymers, 27, 29, 53, 163
ponds, 2, 3, 4, 51

potassium, 61, 92, 97
poultry, 21, 42
precipitation, 15, 42
proteins, 21, 32
pulp, 7, 18, 38, 102
purification, 61, 69, 89, 90, 91, 92

R

raw materials, 103, 105
reactants, 35, 61, 62, 63, 64, 65, 66, 67, 68, 74, 81, 83, 86, 87
reaction time, x, 33, 58, 80, 81, 125
reactions, 59, 61, 62, 74, 95, 96, 160
recovery, 3, 22, 25, 27, 28, 29, 33, 37, 50, 54, 66, 67, 70, 90
recycling, ix, 1, 34, 38, 42, 55
refractive index, ix, xii, 134, 136, 138, 143, 144, 145, 146, 147, 148, 149
remediation, ix, 1, 2, 3, 4, 6, 9, 14, 22, 29, 158, 159, 164
residues, xii, 28, 59, 68, 153
reverse osmosis, 24, 46

S

salinity, 17, 26, 46, 50, 51, 54
sedimentation, 23, 24, 25, 39, 55
seed, xi, 31, 97, 100, 102, 105, 127, 128
simulation, 71, 73, 78, 79, 80, 81, 82, 89, 95, 96
sludge, 3, 15, 16, 29, 37, 42, 45, 49, 162
sodium, 21, 42, 84
solution, 25, 40, 46, 48, 70, 73, 75, 77, 84, 107, 157
solvents, 30, 31, 34, 35, 47, 61, 69, 82
species, ix, x, 1, 2, 3, 4, 5, 7, 9, 10, 11, 12, 13, 15, 16, 17, 19, 20, 22, 26, 29, 32, 46, 49, 57, 60, 71, 72, 76, 81, 158, 159, 160
state, 49, 159
steel, 94
sterile, 5, 12, 20, 53
sulfur, 102, 104, 155, 156
sulfuric acid, 107
surface area, x, 8, 23, 58, 62, 81, 110
sustainability, 9, 28, 30, 49

Sustainable Development, 97
Switzerland, 95
symbiosis, 42, 49
synthesis, x, 21, 39, 40, 49, 58, 69, 76, 80, 82, 84, 88, 89, 91, 92, 93, 94, 95, 96, 97

T

technologies, 6, 46, 59, 61, 67, 90, 91, 93, 95, 96, 158
temperature, xii, 4, 5, 21, 26, 33, 35, 38, 51, 54, 60, 61, 63, 64, 67, 69, 70, 71, 88, 93, 95, 103, 108, 110, 113, 124, 125, 134, 136, 138, 141, 143, 144, 145, 146, 147, 155, 159
ternary blends, 104, 105, 138, 146
toxicity, xii, 153, 155, 156, 157, 159, 160, 161
transesterification, 16, 28, 30, 31, 32, 33, 34, 35, 43, 44, 45, 47, 49, 50, 52, 58, 59, 60, 61, 63, 64, 65, 68, 80, 81, 82, 87, 91, 92, 93, 94, 95, 96, 97, 107
turbulent mixing, 60, 63

U

ultrasound, 36, 61, 63, 82
uniform, 83, 86, 87, 89
urban, 2, 6, 7, 8, 14, 20, 21

V

validation, 83
valorization, 38, 49, 53
vapor, 138, 158
vegetable oil, ix, 31, 59, 69, 87, 91, 150, 156, 163
velocity, 8, 60, 64, 71, 72, 75, 81
viscosity, 60, 71, 87, 102, 103, 112, 115, 117, 119, 124, 125, 136, 147, 149

W

wastewater, vii, ix, 1, 2, 3, 4, 5, 6, 7, 8, 9, 10, 11, 12, 13, 14, 15, 16, 17, 18, 19, 20, 21, 22, 24, 27, 29, 33, 36, 37, 38, 39, 40,

Index

41, 42, 43, 44, 45, 46, 47, 48, 49, 50, 51, 52, 53, 54, 55, 67, 162

water, ix, xi, 1, 5, 9, 22, 30, 32, 33, 34, 35, 53, 59, 66, 68, 69, 99, 122, 129, 163

Y

yeast, 20, 44, 53, 54, 55